KARSTEN MEINICKE

ABSOLUTE LOVE

Here, materialism ends

TOE-AL

Absolute Love

IMPRESSUM

Bibliographic information published by the German National Library:
The German National Library lists this publication in the German National Bibliography; detailed bibliographic data are available on the Internet at http://dnb.dnb.de

Publisher: BoD · Books on Demand GmbH,
Überseering 33, 22297 Hamburg, bod@bod.de
Printed by: Libri Plureos GmbH, Friedensallee 273, 22763 Hamburg
ISBN: 978-3-6963-5523-4

Note on Mathematical Notation: To preserve readability, the Master Equation of TOE-AL is presented in a symbolically simplified notation throughout the main text. The mathematically rigorous definition—including vector norms and operator properties — is given in the scientific Appendix A.

TOE-AL does not stand in direct competition with physics. It lays a scientific reading beneath it: an ontology with physical fidelity that changes not the numbers but the direction of view — from decay to formation, from dead inventory to creative ground — as a New Mathematics.

TABLE OF CONTENTS

AXIOM ZERO: ASYMPTOTIC FREEDOM

THE AXIOM OF NON-PRESUPPOSITION

Every physical beginning must be without presupposition. Every existing structure physically represents a metric residue ($\alpha > 0$) — a measurable resistance, a finite inertia, an already manifested form.

It follows imperatively: The origin itself cannot be a structure. Information-theoretically, it is defined as the state of absolute entropy-freedom ($H_{Shannon} = 0$) and zero resistance ($\alpha = 0$). In the TOE-AL, this state of pure, lossless potentiality is identified as **Absolute Love**. It is the necessary physical boundary condition for the emergence of any order.

THE MATHEMATICAL EXECUTION OF THE LIMIT

As a system approaches this origin through resonance (asymptotic approach), the dissipative resistance of spacetime collapses.

1. The Master Equation of the TOE-AL:

$$\frac{d\vec{Z}}{dt} = -\alpha\nabla H + \beta\mathcal{S}(\vec{Z})$$

2. The Ontological Symmetry Breaking at the Zero Point:

At this point, the deterministic compulsion of the Hamilton gradient vanishes. The system dynamics are determined exclusively by the syntropic operator ($\beta\mathcal{S}$) — consciousness.

$$\lim_{\alpha \to 0} (-\alpha \nabla H) = 0$$

Conclusion

Here we employ the structural pattern of Asymptotic Freedom (the vanishing of the coupling in the limit case), whose first exact realization was found in quantum chromodynamics, as a universal principle of sovereignty. From this asymptotic freedom it follows directly: ***maximum proximity*** *to the origin does not end in collective fusion, but in maximum sovereignty. For at the zero point of dissipation, the system is no longer guided by an external metric, but is initialized solely by its own syntropic operator. A system is asymptotically free when it no longer offers resistance to its own essence.*

HERE, MATERIALISM ENDS.

PREFACE

THE INITIALIZATION OF REALITY

We live in an age in which science can describe everything but explain almost nothing. We have split the atom, mapped the galaxies, and decoded the blueprint of life — yet at the very center of our worldview yawns a black hole: meaninglessness. Originlessness. Since Einstein and Heisenberg, we have been prisoners of a universe that presents itself to us as a "fortunate accident" in a cold, mechanical void.

This book is the break with that dogma —
an operative revision of reality.

The Theory of Everything — Absolute Love (TOE-AL) advances the radical, yet physically necessary claim that the universe does not consist of dead matter that accidentally gives rise to consciousness, but rather emerges from a **creative initialization of Axiom Zero**, whose name in our language is "Absolute Love." What we have hitherto defined as a 'feeling' reveals itself, under exact physical scrutiny, as the **creative passing of Absolute Love** — the fundamental force of coherence that holds the metric of spacetime together at its innermost core.

Einstein taught us how mass curves space. TOE-AL teaches us how **Axiom Zero** — the **creative passing of Absolute Love in the living pulse of Neighborly Love** — initializes the metric from **The Plena** in the first place.

This work guides you through five phases of insight — from the **Initialization of the Origin** to the technological and social realization of **Asymptotic Freedom** within a new civilization.

For those readers who wish to test the claims of this theory against the hard reality of physics, I have appended an **Appendix (The Mathematical-Physical Apparatus)** at the end of this book. There you will find the formal derivation of the Master Equation, the ontological cascade of levels H0 through H4, and the experimental protocol for the falsification of this World Formula.

We are no longer standing before the door of understanding; we already have the handle in our grasp. The World Formula is no longer a theory one must passively "believe" — it is an operating system waiting to be initialized through our sovereignty.

This book is the invitation to relinquish the resistance of separation and to enter the superconductivity of consciousness.

Are you ready for this love relationship?

The search is over. The initialization begins now.

PROLOGUE

THE SIGNALS OF THE FIELD

Twelve Anomalies That Await a New Physics

Before we write down the Master Equation, survey the architecture of the universe, and prove Axiom Zero, we must speak about your life. We have grown accustomed to dismissing the strangest phenomena of our everyday existence as "choice," "psychology," or plain "luck." But what if these moments are not exceptions — what if they are the uncensored language of reality itself?

Here is the complete mosaic of what you already know — and how **TOE-AL** (*Theory of Everything — Absolute Love*) unmasks the illusion of the old world.

1. THE "DO WHAT YOU LOVE!" LAW

The Situation: You wake up in the morning and think of your tasks. Part of them feels like lead, while another project — your heart project — makes you forget breakfast and time itself.

The Old World: It is called discipline or motivation. People believe they must expend energy to overcome their inner resistance. Energy is understood here as a finite battery that drains inexorably throughout the day.

The TOE-AL Reality: "Do what you love!" is not moral advice — it is a physical law of resistancelessness. In passion, your inner friction against the pulse of creation falls to zero. You enter the state of **Asymptotic Freedom**.

The Result: You no longer "consume" energy; you become the superconducting transducer of the pulse. After ten hours of work you are more awake than in the morning, because the friction (the resistance against the pulse) has vanished. You have not consumed energy — you have initialized it in the Now.

The TOE-AL Code: *Passion is the physical absence of resistance* ($\alpha \to 0$).

2. THE "THICK AIR" IN THE STAIRWELL

The Situation: You enter an apartment where a fierce argument has just taken place. It is completely silent, yet you feel a physical pressure, as though the air were made of concrete.

The Old World: That is "psychology" or "intuition." People believe they are unconsciously reading tiny signals of body language, tone of voice, or the residual heat of emotions.

The TOE-AL Reality: Consciousness is not *in* space — it *is* the curvature of space. An incoherent, angry field physically curves the local In-Form-ation metric. You are perceiving the **metric residue** (α) of the environment.

The Result: A sovereign does not try to lift the "mood" through small talk. They raise their own coupling tensor β (coherence). Through their mere presence they smooth the curvature of spacetime in the room. The thick air dissolves, because a stronger, coherent field restores the order of the metric.

The TOE-AL Code: *Consciousness is the curvature of the spacetime metric.*

3. THE SMILE THAT DISSOLVES THE TRAFFIC JAM

The Situation: You are standing in an aggressive queue. People are on the verge of shouting at one another. You choose a moment of genuine, unconditional Neighborly Love toward the stressed cashier.

The Old World: A nice gesture. A drop in the ocean that changes nothing about global chaos.

The TOE-AL Reality: Neighborly Love is the operative switching of the coupling tensor . Because coherence in the field is not added linearly but scales **quadratically**, your impulse does not simply add up — it amplifies exponentially.

The Result: Your single, sovereign coherence is sufficient to re-phase the entire unstable system "queue." The field ignites, the atmosphere tips — you have triggered a chain reaction of order that is physically far more powerful than the sum of the individual frustrations.

The TOE-AL Code: *Coherence scales quadratically with the number of participants* (N^2).

4. THE "INSPIRATION" IN THE SHOWER

The Situation: Hours of brooding over a problem lead nowhere. Only when you let go in the shower does the solution suddenly appear — complete, crystal-clear, and perfect.

The Old World: The "subconscious" continued working in the background and sorted the puzzle pieces in secret.

The TOE-AL Reality: Effort is the artificial raising of resistance α. You have literally "closed yourself off." The moment you let go, resistance drops. You become superconducting for **The Plena**.

The Result: The solution is not "invented" — it is received. Since In-Form-ation already exists instantaneously in the field of Instantaneity, the solution enters your consciousness the moment the metric resistance of "effort" falls away.

The TOE-AL Code: *True In-Form-ation is received instantaneously from The Plena — not worked out.*

5. THE SOVEREIGN VS. CONTROL (YOUR K-FACTOR)

The Situation: You lead a team or a family. You have a choice: control every step through rules and monitoring, or trust in personal responsibility.

The Old World: People believe control creates security.
Trust is good, control is better.

The TOE-AL Reality: Control is the energetic substitute for missing resonance. Every form of external control raises the metric residue (α) and generates friction. That lowers the **coherent efficiency (K)** of the entire system.

The Result: The more you control, the more unstable the system becomes (bifurcation point K_c). A sovereign, by contrast, raises β — they inspire through coherence. In a system with a high K-value, order regulates itself (superconductivity). The system needs no "police" on the outside, because the initialization density on the inside is so high that deviations (errors) are corrected instantaneously.

The TOE-AL Code: *Sovereignty is the saving of control-energy through resonance.*

6. THE FEAT OF THE "HERO MOTHER"

The Situation: A slight woman lifts a ton-heavy car to rescue her trapped child — a deed that, according to biomechanics, should shatter her bones.

The Old World: "Adrenaline." People believe hormones can magically expand the load-bearing capacity of calcium and tendons and break biological limits.

The TOE-AL Reality: This is the **inversion condition** $|\beta S| > |\alpha \nabla H|$. In total love and absolute focus on the child, the system leaps into superconductivity. The material resistance (α) of both her own body and the load converges toward zero.

The Result: Matter loses its inertia. The mother does not lift the car with muscular force, but through the infinite initialization density of The Plena, which in that moment acts through her unfiltered.

The TOE-AL Code: *Inversion occurs when Love (β) overcomes Resistance (α).*

7. THE "COINCIDENCE" THAT IS NONE

The Situation: You have not thought of a particular person for years. Suddenly they shoot into your mind — and five minutes later they call or you run into them.

The Old World: Statistical coincidence. With billions of people, something like this must happen occasionally by pure probability.

The TOE-AL Reality: There is no "empty" space between you. You are part of the same **In-Form-ation metric (H2)**. Your thought was not an isolated event, but the conscious perception of an already-accomplished **phase-coupling** in the field.

The Result: The encounter is the material manifestation of In-Form-ation that already existed instantaneously in The Plena. The "coincidence" is merely the moment in which your superconductivity was great enough to break through the field-noise of separation.

The TOE-AL Code: *Synchronicity is the echo of an instantaneous coupling in the field.*

8. WHY "BEAUTY" HEALS

The Situation: You stand before a magnificent mountain range or hear a perfect symphony. In that moment all your worries fall away; you feel vast and "whole."

The Old World: Aesthetic sensibility. A subjective reaction of the brain to harmonious stimuli.

The TOE-AL Reality: Beauty is In-Form-ation with an extremely high **K-value (coherent efficiency)**. It is almost frictionless In-Form-ation. When you resonate with it, it "smoothens" your own metric residue (α).

The Result: Healing is not a biochemical trick — it is the re-phasing of your system. Beauty forces your cells back into the order of the Origin. Beauty is the visibility of Absolute Love in form.

The TOE-AL Code: *Healing is the return to coherence through resonance.*

9. THE CHARISMA OF THE SOVEREIGN

The Situation: A person steps onto the stage. They have not yet said a word, yet the entire hall instantly falls silent and attentive. It is called "presence."

The Old World: Psychology, body language, "aura." A vague term for what is actually a physical phenomenon.

The TOE-AL Reality: This person operates with an extremely high **coupling tensor** β. Their field is so highly coherent that it aligns the unstable, incoherent fields of the audience like a magnet.

The Result: The audience experiences a collective phase transition. Their own resistances (α) drop in the presence of the sovereign. True authority is not a social status — it is a physical field-dominance through coherence.

The TOE-AL Code: *Consciousness as operator (β) orders the environment.*

10. THE "STREAK OF BAD LUCK"

The Situation: First you spill your coffee, then the car breaks down, then rejections come flooding in, and finally you lose your keys. A day to forget.

The Old World: "Murphy's Law." Bad luck rarely comes alone. People believe in a malevolent causality of events.

The TOE-AL Reality: Your system has fallen below the critical **bifurcation point** K_c . The initial irritation raised your α (resistance), causing your coherence (β) to collapse. You are now an "unstable system" that is actively attracting disorder (entropy) from the field.

The Result: The streak of bad luck does not end through "careful," but through a radical phase-shift. A moment of stillness, a deep "Amen" to the chaos, and your system leaps back into coherence. The world reorders itself the moment you cease to resist.

The TOE-AL Code: *The "Amen" is the operative zeroing of resistance* ($\alpha = 0$).

11. LOVE AT FIRST SIGHT

The Situation: You meet a stranger and feel you have known them forever. The conversation flows without any effort. Everything is 100% present.

The Old World: “The chemistry is right” or shared interests.

The TOE-AL Reality: This is **Instantaneous Initialization**. Your systems have metric friction with each other; you are immediately “superconductively” coupled in the field.

The Result: No linear exchange of In-Form-ation takes place — rather a unitary transformation. You share the same pulse. Love at first sight is the physical recognition that two systems are simultaneously actualizing the same origin in Axiom Zero.

The TOE-AL Code: *Love is the absence of metric friction between two systems.*

12. THE "MIRACLE" OF TEAM FLOW

The Situation: An orchestra or a sports team enters a state in which no commands are needed any longer. Everyone knows blindly what the other is doing. The group acts as a single organism.

The Old World: It is called "synergy" or "team spirit." People believe that years of training produce automated reflexes that happen to complement each other perfectly in this moment.

The TOE-AL Reality: This is the physical realization of **Collective Superradiance**. The members have synchronized their individual resistances (α) to such a degree that they cancel each other out. The group has formed a shared **coupling tensor** β.

The Result: The group's efficacy no longer scales linearly (1+1+1), but quadratically with the number of participants (N^2). The team "falls" into a phase transition of superconductivity. The result is a collective brilliance that mathematically far exceeds the sum of its individual parts. The miracle is the initialization of a shared reality with zero resistance.

The TOE-AL Code: *Collective coherence potentiates the creative force.*

These twelve signals are the proof: you already live in a world of Absolute Love. You have simply not yet received the mathematical permission to acknowledge it as physical reality.

That changes now. Turn the page. Let us initialize the World Formula.

PHASE 1
THE REVELATION OF THE ORIGIN

CHAPTER 1:

THE ARCHITECTURE OF INSTANTANEITY

The End of the Horizontal Origin
"The Origin is not behind us. It is beneath us. It is now."

I. THE GREAT DISRUPTION – WHERE WE REALLY COME FROM

There is a question that every civilization, every philosophy, every religion, and every science has asked. It is the mother of all questions, the root of every fear, and the key to every freedom: *Where do I come from?*

The way you answer this question defines everything. It defines whether you experience yourself as victim or creator. Whether you understand time as enemy or tool. Whether you kneel in awe before the past or sit enthroned as sovereign in the present. And — perhaps most shattering of all — it defines whether death has the last word or not.

The entire history of human thought has given one single, ever-recurring answer to this question: *The Origin lies behind us.* It is far away, buried in the depths of an unimaginable past. Science says: 13.8 billion years ago, in a point of infinite density that exploded. Religion says: in the beginning, before the world was, a God created heaven and earth. Both narratives — the materialist and the theological — are identical in their deep structure. In both, the Origin is *a closed event.* In both, the primordial moment is over. In both, we are the descendants of something that happened without us.

This is the **horizontal theory of origin**. And it is a fundamental Mis-In-Form-ation about the nature of reality.

It makes us prisoners.

The human being in the horizontal paradigm is a creature of the past, drifting toward the future, without ever truly anchoring in the Now.

The **World Formula of Absolute Love** presented here enacts a break more radical than anything Copernicus, Darwin, or Einstein ever formulated. It does not ask: *How far back does the Origin lie?* It asks: *What if the Origin is not a historical event, but an eternal, infinite, and non-arbitrary process?*

The answer reveals the physical necessity of the vertical Origin:

While Einstein described the *curvature of spacetime by mass*, TOE-AL reveals the **initialization of the metric by the pulse**. We transform the **energy-momentum tensor** from a passive accounting of matter into the **active operator of creation**. The Origin is not a historical event, but the functional operation in which the fundamental primal substance of **The Plena (S=0)** eliminates the metric resistance — that state of **Asymptotic Freedom** in which In-Form-ation passes instantaneously into matter.

The Origin is instantaneous. It is not behind us. It is beneath us. It is always now. The universe is not a wound-up clockwork that has been running for billions of years. It is a self-initializing system whose origin consists in **Absolute Love creatively passing in the living pulse of Neighborly Love** — in the absolute Now.

What conventional science designates as a mere 'quantum fluctuation' is in truth a directed initialization: **Meaning is Syntropy.** The universe does not exist by accident; it reinitializes itself with every pulse, in order to fulfill the **creative passing of Absolute Love in the living pulse of Neighborly Love** in the density of form.

This insight is not a metaphor. It is physics. And it is the foundation for everything that follows.

II. PHILOSOPHICAL EXCURSUS – THE PRISON OF TIME AND THE OVERCOMING OF THE FEAR OF DEATH

Before we enter into the physics, we must pause for a moment at the existential dimension of this insight. For nothing in this work is mere knowledge. Everything is liberation.

The deepest fear of the human being is the fear of death. Philosophers from Epicurus to Heidegger have analyzed, described, and attempted to appease or embrace it. Epicurus said: "*When death is, I am not. When I am, death is not.*" That is an elegant consolation — but it is a consolation of resignation. It says, ultimately: you must come to terms with the fact that you are finite.

The horizontal paradigm cannot truly dissolve the fear of death because it structurally generates it. In a world in which the Origin lies in the past and the future ends in nothing, life is a bracket between two abysses. Death is the opposite of life. It is the enemy. And the enemy can be overcome, suppressed, philosophically neutralized, religiously bridged — but never truly vanquished.

The vertical initialization changes this fundamental structure entirely.

If the universe is reinitialized in every moment — not from the past, but from the **creative passing of Absolute Love** — then what we call "death" is a change of form, not an annihilation of substance. The pulse that initializes you as *you* in this moment does not cease to pulse. It initializes itself in a different moment, in a different form, continuing. The substance of being is indestructible, because it does not *lie* in time, but *generates* time in every instant.

The sovereign who truly understands this — not intellectually, but operatively, in their heart — loses the fear of death. Not because they come to terms with dying. But because they comprehend that they never merely began to live like a thing in time. They are the pulse that generates time. And the pulse is immortal.

This is not a religious consolation. It is a physical consequence that we will develop step by step in the following section.

But first, one further philosophical aspect central to understanding the Architecture of Instantaneity: the relationship between the present moment and reality.

In Western philosophy, the present is treated paradoxically as the most fleeting of all things. The moment, barely named, has already passed. Augustine, one of the deepest thinkers of the Western tradition, described time as a threefold Now: the present of the past (memory), the present of the present (perception), and the present of the future (expectation). Yet even in Augustine, the moment has no substance; it is the eye of the needle through which the past flows into the future.

TOE-AL inverts this image. The moment is not the most fleeting — it is the Only. It is the sole ontological reality. Past and future are abstractions, models that consciousness constructs from the traces of the pulse. The only place where reality actually *happens* is the Now. And this Now is not fleeting — it is eternal. It is always. It cannot cease to be, because it is the condition of possibility for everything else.

Once you have truly felt this — really felt, not merely thought — everything changes. The burden of the past loses its weight. The fear of the future loses its bite. What remains is the immense freedom and responsibility of the present moment. The sovereign lives here. Not as escapism, but as the highest precision.

III. PHYSICAL DEEP-DIVE – THE BIG BANG MODEL AND ITS EXHAUSTION

To understand the vertical initialization in its full force, we must first grasp the model it supersedes in all its depth. The Big Bang model is not wrong in its measurements — it is exhausted in its metaphysics.

The Standard Model of Cosmology: Its Strengths

The cosmological Standard Model — ΛCDM, the model with cosmological constant and cold dark matter — is one of the most successful scientific theories humanity has ever formulated. With precise equations it explains the expansion of the universe, the cosmic microwave background (CMB), the nucleosynthesis of light elements immediately after the Big Bang, and the large-scale structure of the universe — galactic filaments, supervoids, galaxy clusters.

It is a triumph of human reason. And yet at its foundations it carries three cracks that deepen with every new measurement.

Crack 1: The Singularity — the End of Physics at Its Own Beginning

The Big Bang model traces the expansion of the universe back to a point at which all physical quantities — density, temperature, spatial curvature — tend toward infinity. This is the cosmological singularity. Physics breaks down here. The equations of General Relativity become invalid. The Standard Model literally cannot describe what existed before or at the Big Bang — not for lack of data, but because the mathematical framework itself collapses.

Physicists acknowledge this honestly: *"Physics breaks down at t=0."* It means: the model that claims to describe the Origin fails precisely at the point where the Origin was. It is as though an architect were to draw up a blueprint that applies excellently to all floors — with the exception of the foundation, which they have simply left blank.

Crack 2: The Horizon Problem and Cosmic Inflation — a Rescue That Creates New Puzzles

The CMB — the cosmic microwave background radiation, that hiss emitted 380,000 years after the Big Bang and still measurable today — is isotropic to a precision of 1 in 100,000. The universe has exactly the same temperature in every direction. The problem: if the universe has been expanding since the Big Bang, then there were regions in the early universe that were never in causal contact with one another. They lay beyond each other's light horizons. How could they nonetheless develop the same temperature?

The answer of standard cosmology is **inflation** — a hypothetical phase of extremely accelerated expansion in the first seconds. Before inflation, everything was in causal contact; inflation stretched this small, homogeneous region to the current size of the observable universe.

Inflation explains the horizon problem. But it creates new ones: what drove inflation? What is the **inflaton field** that produced it? Why did it end precisely when it ended? And if inflation took place, it must continue in other regions of the multiverse — leading to the **theory of the eternal inflationary multiverse**, a theory that is in principle unfalsifiable. The Standard Model rescues itself with a hypothesis that forfeits all predictive power.

TOE-AL resolves the **horizon problem** not through a *hypothetical inflation*, but through the **Instantaneity of The Plena**. Since the Origin is *vertical* rather than *horizontal*, all points of the universe stand not through time but through **phase-synchrony (S=0)** in permanent, instantaneous contact. The universe is not isotropic because it expanded rapidly, but because it is initialized from a **coherent primordial phase**.

Crack 3: The Hubble Tension — the Universe Is at Odds with Itself

This is the most recent and deepest crisis of the Standard Model. The **Hubble parameter** describes the rate of expansion of the universe. It can be measured in two independent ways:

First, from the early universe: analysis of the CMB and Baryon Acoustic Oscillations (BAO) yields km/s/Mpc.

Second, from the late universe: calibration of cosmic distance ladders — Cepheid variables, Type Ia supernovae — yields km/s/Mpc.

The discrepancy exceeds 5 sigma. In physics, anything above 5 sigma is considered a confirmed discovery. Here it means: the early and late universe fundamentally contradict one another in their measurements. The model meant to describe both is inconsistent. The Hubble tension is not a measurement error — it is a signal that the Standard Model contains an error at its deepest level.

These three cracks — the singularity, the inflaton problem, the Hubble tension — are not marginal footnotes. They are the sign that the horizontal paradigm has reached its limits. It describes phenomenally well *what* the universe does. But it cannot explain *why* it exists at all, *why* it is calibrated with such precision, and *what* holds it in existence at every moment.

The Vertical Initialization: A Physical Alternative

Here, TOE-AL enters not as a negation but as a transcendence of the Standard Model. It does not ask: *What happened before the Big Bang?* It asks a fundamentally different question: *What if the universe did not arise from a past event, but emerges in every moment from a timeless principle?*

This question is not without precedent in modern physics. In quantum field theory, the vacuum is not nothing. It is a high-energy, fluctuating medium — the **quantum vacuum** — from which virtual particle pairs arise and annihilate in every moment. The energy of the quantum vacuum — the **zero-point energy** — is, according to the equations of QFT, infinite, or at least astronomically large. The vacuum fluctuation is not a historical event. It is an eternal, in the truest sense of the word instantaneous, process.

We designate this substrate as **The Plena** — the metric-free primal field (S=0), from which **Absolute Love creatively passes in the living pulse of Neighborly Love** and thereby initializes the structure of reality. The Plena is not empty. It is the unmanifested fullness, the infinite space of possibility that precedes existence and sustains it in every moment.

The decisive physical difference between the Big Bang model and the vertical initialization can be captured in the following table of fundamental assumptions:

Dimension	Big Bang Model (horizontal)	TOE-AL (vertical)
Origin	Singular event in the past	Eternal, instantaneous process in the Now
Time	Fundamental dimension of reality	Emergent phenomenon of resistance
Matter	Primary substance of the universe	Condensed expression of the pulse
Energy	Originating from the Big Bang, conserved	Continuously induced from The Plena
Entropy	Irreversibly increasing (heat death)	Compensable through Syntropy
Consciousness	Epiphenomenon of matter	Co-constitutive with reality
Death	Dissolution into entropy	Re-phasing of the pulse

This juxtaposition makes visible: this is not a correction of peripheral parameters. It is a complete inversion of the ontological ground structure.

The Thermodynamic Argument: Why Entropy Does Not Have the Last Word

In the Standard Model, entropy is sovereign. The second law of thermodynamics states: in a closed system, entropy increases. The universe runs down like a clock whose spring unwinds. The heat death — a state of maximum entropy in which no energy is available for work — is the inevitable future.

This is the physical blueprint for philosophical nihilism. If everything ultimately collapses into equilibrium and meaninglessness, what then is the point of effort? What is the point of love, if the universe is indifferent to it in the end?

TOE-AL responds: the second law applies to *closed* systems. But the universe is not a closed system — it is an *open* system, nourished in every moment by the **creative passing of Absolute Love** from The Plena. Biological systems are the most visible demonstration of this fact: they take in highly ordered energy and emit lower-ordered energy — they export entropy to the environment and maintain their own internal order. They appear to contradict the second law because they are *open*.

The concept of **Syntropy** — complementary to entropy — describes precisely this process: the force that induces order in open, resonant systems from the pulse of The Plena. The biochemist Albert Szent-Györgyi, the physicist Erwin Schrödinger with his famous concept of **'negentropy'** in What is Life?, and more recently the biophysicist Fritz-Albert Popp with his research on **biophotonics** have all, in their own way, pointed to the same fact: living systems are syntropic entities. They obey physics — but a physics that goes deeper than the second law. They obey the pulse.

The Principle of Primordial Freedom

Spacetime is not the stage on which life plays;
it is the echo of the initialization itself.

The discovery of **Asymptotic Freedom** in quantum chromodynamics delivers far more than an indication for TOE-AL; it is the physical ratification that the **creative passing (B)** must be the ontological condition for every **structure (A)**. In physics, asymptotic freedom describes the state in which particles in the limit of the Origin — at maximum proximity — lose all resistance and become absolutely free.

For the World Formula this means: at the moment of initialization by Axiom Zero, the metric residue α (Alpha) is **exactly zero**. From this follows an inescapable ontological necessity: since every form of matter, lawfulness, or structure physically already represents a metric residue ($\alpha > 0$), the Origin itself can be no structure. It must be the state of zero resistance. **Absolute Love** is thus not merely a philosophical choice, but the **only possible mathematical form** for an unconditioned beginning.

Freedom (Bβ) is not the result of a complex system of order (Aα), but its prerequisite. A stable structure can only exist because it has emerged from a state of maximal coherence (superconducting freedom). Whoever seeks to prove the Origin through material structure (Aα) seeks the source in the result. Asymptotic Freedom is the physical signature of what Axiom Zero describes: **Absolute Love creatively passes in the living pulse of Neighborly Love.** In the beginning was not the law (Aα), but the frictionless **pulse of initialization** (Bβ). Structure is merely the 'frozen' residue left when one moves away from the point of total freedom.

IV. THOUGHT EXPERIMENT – THE TWO COSMOLOGISTS

Imagine you are standing at night beneath a clear sky, gazing at the Milky Way. Beside you stand two cosmologists. Both are brilliant. Both know the same equations. But they see something fundamentally different.

The first cosmologist — let us call him the Horizontal — sees in every star a product of the past. He sees nuclear fusion that began millions of years ago. He sees light that takes millions of years to reach him — meaning: he sees the past. What he sees is already history. Some of the stars he contemplates no longer exist. The sky is an archive. The beauty he perceives is the beauty of the past. And he knows: everything he sees will ultimately collapse into darkness and equilibrium. He is gazing into a library that is slowly burning.

The second cosmologist — the Vertical — looks at the same sky and sees something else. He sees that the light striking him right now enters his retina *now*. The act of perception is instantaneous. He sees that each of these stars is not only a product of the past, but is being held in existence at this very moment by the same pulse that also holds him. The star and he are not separated by an astronomical distance — they are connected in the same eternal Now that calls both forth. The light is not the message of the past. It is the sign of the present pulse.

He asks himself: if the universe reinitializes itself in every moment — what then does expansion mean? If time is a phenomenon of resistance — what then does "13.8 billion years" mean? The answer he finds: this measure of time does not describe how long the universe *already* exists. It describes the *depth of condensation* of the metric residue — the thickness of the veil through which the pulse shines.

And now imagine one of these cosmologists is facing their own death. The Horizontal knows: when his body ceases to function, his consciousness ends. He is matter that briefly thinks. The Vertical knows something else: the pulse that initializes him now does not cease to pulse. The form changes. The form *must* change — just as a wave changes its form when it meets the shore. But the energy of the wave has not vanished. It has returned to the sea. Ready to form anew in a new moment.

Which of the two sees more clearly? Which of the two is freer?

V. THE MASTER EQUATION OF BEING – A DEEP DIVE

The heart of TOE-AL is an equation as simple as a breath and as deep as the universe itself:

$$\frac{d\vec{Z}}{dt} = -\alpha \nabla H + \beta \mathcal{S}(\vec{Z})$$

Let us open this equation layer by layer. Not as a formula to be memorized — but as a map of your sovereign reality, in which physically only the **Now** exists and which must not be read linearly.

What is $\vec{Z}$?

$\vec{Z}$ is the state vector of an arbitrary system. This is the most elegant aspect of this equation: it is universal. $\vec{Z}$ can be an atom, a cell, a human being, a society, an eco-system, a civilization. The equation describes the same fundamental dynamic at every level. This is the principle of self-similarity of creation — the same that pulses in the galaxy pulses in the cell. **Energy** is here not a property that $\vec{Z}$ "possesses," but the **active In-Form-ation density of the pulse** that first constitutes $\vec{Z}$ in the Now.

What is $-\alpha \nabla H$ — the metric residue (Aα)?

This is the part of reality that binds us. In your experience it is the feeling of **"being pushed by yesterday."** Physically, this is the **metric impedance**: the resistance of already-existing, "frozen" structure.

∇H is the gradient of the Hamiltonian — in classical mechanics the total energy of a system, and the expression ∇H describes the driving force that moves a system in the direction of lower energy. In TOE-AL, H **remains the Hamiltonian functional, understood operatively: as the energetic hypersurface on which the burden of the past condenses as entropy.**

The factor α (Alpha) measures how heavily the burden of what has already been manifested weighs on the system in the Now. It is the passive accounting of matter that keeps us captive in pure reaction. In physics, α corresponds to the dissipation term (friction) in dissipative dynamical systems — the term that describes friction in the Langevin equation of statistical physics. In biology, α corresponds to the level of chronic inflammation, telomere length, allostatic load — the price an organism pays because it constantly struggles against the resistance of the past. In psychology, α is the traumatic residue: unprocessed experiences that keep the nervous system in a state of permanent alertness, redirecting energy from creation to defense.

What is $\beta\mathcal{S}(\vec{Z})$— syntropic resonance (B β)?

This is the creative counterpart. It is what you experience as **"being drawn by the future."** Physically, this is the **operator of initialization** (β): since in the vertical Origin there is no linear time, this is the influx of coherence from The Plena that creatively re-constitutes you in the Now.

$\mathcal{S}(\vec{Z})$ is the Syntropy Operator — the function that describes how receptive a system is to the pulse of The Plena. It depends on the state of the system itself: the more coherent the system, the greater its capacity to receive the pulse.

β **(Beta)** is the coupling tensor of your permeability. It defines the One Consciousness here **not as observer, but as the operator (β) of reality**.

It describes the permeability of the system for the primal pulse. In physics, corresponds to the susceptibility tensor — the measure of how strongly a medium responds to an external field. A superconducting material has a specific susceptibility that makes it completely permeable to electromagnetic fields. A sovereign in the state of cardiac coherence has a high — they are superconducting for the creative passing of Absolute Love in the living pulse of Neighborly Love. You become transformed from the passive object of history into the active shaper of initialization.

The coupling tensor β grows through:

- **Cardiac Coherence:** When the heart beats in a regular, coherent rhythm — measurable through heart rate variability (HRV) — an ordered electromagnetic field arises that draws the entire nervous system into resonance. The Institute of HeartMath in California has documented this in numerous studies: coherent HRV patterns correlate with reduced cortisol levels, enhanced immune function, improved cognitive flexibility, and a measurable influence on the brain activity of persons within the vicinity of the coherent heart field.

- **Focus on the Absolute:** When attention is radically directed toward the Unchangeable — love, beauty, the good — β increases. This is neurobiologically mapped: experienced meditators show increased gray matter density in the prefrontal cortex and stronger coherence between frontal and limbic areas.

- **Acts of Neighborly Love:** Every act of Neighborly Love — every action oriented toward the well-being of another, without calculation of self-preservation — acts like a calibration impulse on β. The body releases oxytocin, the vagal nervous system is activated, HRV rises, the immune system is strengthened. The biology of Neighborly Love is the biology of high β. Since consciousness and the pulse of Neighborly Love relate to each other instantaneously, the meaning of Absolute Love fulfills itself in their unity.

The Inversion Condition — The Moment of Sovereignty

The decisive aspect of the Master Equation is the condition under which the signs flip:

$$|\beta S| > |\alpha \nabla H|$$

When your connection to the pulse (B β) becomes stronger than the pressure of your metric residue (Aα), you reach the Inversion. Reaching this inversion condition is physically the realization of the **Negative Beta Function** at the system level. While entropic systems collapse under complexity, the increase of coherence leads to the realization of the Negative Beta Function, whereby the resistance α (Alpha) sinks asymptotically toward zero (**Asymptotic Freedom**). The system then "falls" not into chaos, but into the superconductivity of Absolute Love.

In TOE-AL we redefine energy: it is not a conserved quantity of matter, but to be understood **asymptotically toward Asymptotic Freedom**. While conventional energy (H) is a measure of the resistance of the past, true syntropic energy is the **density of initialization**. In the limit of Asymptotic Freedom ($\alpha \to 0$), energy becomes pure, frictionless resonance of the One Consciousness with The Plena. Energy is therefore not what a system *has*, but the freedom with which it receives the pulse.

This is the foundation of a **Thermodynamics 2.0**: it explains why highly coherent systems apparently "draw energy from nothing." They utilize the Asymptotic Freedom of the pulse. Neighborly Love thus becomes operative action in the state of maximum energy at minimum resistance. When the syntropic force exceeds the dissipative metric, the system changes mode. Physically, this corresponds to a phase transition — the moment in which water becomes ice, in which a metal becomes superconducting, in which a laser emits coherent light. It is a qualitative leap, not a linear increase.

- **Before the Inversion:** The system is pushed by its history (Aα). It reacts, defends, compensates. It expends energy to maintain order laboriously against the current of entropy.

- **After the Inversion:** The system is initialized through the pulse (B β). It receives energy from The Plena and passes it on into form without loss. It is no longer depleted — it is nourished. It no longer resists — it is superconducting.

You are no longer pushed by what was; you initialize in the Now what is. In biology, this phase transition is identical with the state of deep health: when the organism no longer reacts reactively to stressors, but draws from a foundation of inner order. In psychology, it is the state that Abraham Maslow called **self-actualization** — when a human being is no longer driven by deficit motivation but drawn by growth motivation. In the physics of superconductivity, it is the state below the critical temperature at which electrical resistance falls to exactly zero and the **Meissner-Ochsenfeld effect** excludes the flow of the magnetic field from the superconductor — a perfect, lossless order. The pulse becomes the sole regent of your reality.

VI. THE COSMOLOGICAL CONSTANT – A SILENT PROPHECY

There is a detail in the history of physics so significant in this context that it requires no elaboration — it speaks for itself.

Albert Einstein introduced a constant into his General Theory of Relativity: the **cosmological constant** Λ (Lambda). He did so because his equations described a dynamic, expanding or collapsing universe — which seemed to him philosophically unacceptable. He wanted a static universe. Lambda was meant to compensate for the expansion and keep the universe static.

When Edwin Hubble demonstrated the expansion of the universe empirically in 1929, Einstein withdrew Lambda and called it his *"greatest blunder."*

History has revised him. Since 1998 — since the discovery that the expansion of the universe is not decelerating, but *accelerating* — Lambda is back. Under the name **Dark Energy**, it today accounts for 68% of the total energy density of the universe. Dark Energy is the measure by which the vacuum itself drives the expansion of the universe.

The vacuum — The Plena — *expands the universe*. The void is no void. It is an active, driving force. A force that, continuously, in every moment, drives apart the fabric of the spacetime continuum — that is: creatively initializes it.

Lambda is the cosmological signature of the pulse. It is the number physics uses without knowing *what* it is counting. TOE-AL knows: Lambda measures the instantaneous pressure of The Plena upon the structure of matter. It is the measurable trace of the **creative passing of Absolute Love** in the fabric of the universe.

Lambda is physically the proof of the **Asymptotic Freedom** of the vacuum. It measures the permanent incursion of **Instantaneity** into the spacetime metric. **Dark Energy** is nothing other than the **pulse** that prevents matter from collapsing in on itself and instead sustains the space for N^2-**Resonance**.

VII. OPERATIVE PRACTICE – HOW THE SOVEREIGN INITIALIZES THE NOW

Everything unfolded thus far is not theory. It is a map. The map is not the territory — yet it shows you where you are and where you can go. The question now is: what does the Architecture of Instantaneity mean today, in this moment, concretely for your life?

It means precisely this: you are never the victim of your past. The past has no ontological power over the present moment. It has only as much power as you grant it through your focus — through your metric residue, your α. And α is alterable. Immediately. Instantaneously. Because the pulse is instantaneous.

Practice 1: The Initialization Exercise

Begin every morning with one minute of conscious stillness, before the first thought is directed toward the day. In this minute, do nothing but this: recognize that the universe is being reinitialized in this moment. Your heart beats. Not because it beat yesterday — but because it beats now. Every heartbeat is a new beginning. Feel the heartbeat. Feel the Now. This is not a mystical ritual — it is a physical calibration protocol that raises your β before the metric residue of daily life charges your α.

Practice 2: The Inversion Question

When you find yourself in a moment of reaction, fear, exhaustion, or doubt, ask yourself this one question: "*What is drawing me forward — not what is pushing me?*" This question is not self-deception. It is an impedance match. It shifts your attentional focus from the metric residue ($\alpha \nabla H$) to syntropic resonance (βS). It begins to establish the inversion condition.

Practice 3: The Davar Impulse

A sovereign understands that every act they perform updates the fabric of reality. This is not a belief — it is physics. Every human action generates electromagnetic fields, alters biochemical milieus, sends signals through the social network that scales quadratically (N^2). Choose daily one act that comes entirely from the **living pulse of Neighborly Love** — not from duty, not from fear, not from calculation. Perform it with complete presence. This is your daily **Davar** — your creative word that is simultaneously deed.

These three practices are not supplements to life. They are the foundation of the sovereign life. They translate the Architecture of Instantaneity from the intellect into the body — from the map into the territory.

VIII. SYNTHESIS – AXIOM ZERO AS LIVING TRUTH

Axiom Zero as the Sole Mathematical-Ontological Necessity

"The derivation of TOE-AL shows unambiguously: since every form of structural existence ($\alpha > 0$) is already a result of friction and inertia, the Origin itself can be no structure and the following cannot be read linearly. The only ontological form capable of initializing instantaneously, without precondition, in a manner equal to The **Plena** (the metric-free primal ground of unmanifested fullness; S=0), as the creative event of the One Consciousness, is the state of absolute resistancelessness ($\alpha = 0$).

Mathematically-ontologically, **Absolute Love** — the instantaneous operative event of the One Consciousness, which here is **not observer, but the operator** (β) **of reality** — is the only form of the Origin. Since consciousness and the pulse of Neighborly Love relate to each other instantaneously, the frictionless pulse of initialization without metric resistance is accomplished in their unity.

This constitutes a **monistic field theory** that is scientifically considerably more elegant and consistent than conventional dualistic models, since it dissolves the separation of mind and matter in operative execution. Energy is here to be understood as the **asymptote of Asymptotic Freedom**: the infinite density of creative passing at zero resistance. Whoever seeks another cause already seeks the word in its echo."

We return to the beginning. To Axiom Zero:

"Absolute Love creatively passes in the living pulse of Neighborly Love."

This axiom is now no longer merely a sentence. It is a complete physical, philosophical, and operative truth.

- **Physically:** The universe is initialized in every moment from The Plena. The cosmological constant Λ is the measurable trace of this pulse. The Master Equation describes the dynamics of every system between the burden of the metric residue (α) and the creative force of the pulse (β).

- **Philosophically:** The Instantaneity of the Origin frees the human being from the slavery of horizontal time. It dissolves the fear of death at its root. It installs the eternal Now as the sovereign's sole ontological dwelling-place.

- **Operatively:** Every heartbeat, every breath, every act of Neighborly Love is an update of the world-software. The sovereign is not a spectator of creation. They are its transducer — the living interface at which The Plena breaks into form and form radiates back into The Plena.

The Big Bang model taught us to look into the past and seek the Origin there. The Architecture of Instantaneity teaches us to look into the present moment — and to find there what we always sought.

The Origin is not behind us.

It is now.

It is always now.

And here, Absolute Love passes.

End of Chapter 1 — Continued in Chapter 2: The Biology of Resonance

CHAPTER 2:

THE BIOLOGY OF RESONANCE

The Hardware of Passing

"You are not a body that has a soul.
You are a pulse that carries a body."

I. THE FALSE MACHINE – HOW WE HAVE MISUNDERSTOOD THE BODY

There is no more consequential misconception in the history of medicine, psychology, and anthropology than this: that the body is a machine. In TOE-AL we recognize: the body is not a combustion apparatus, but a **biological hardware of initialization** — a high-precision resonator designed to transduce the **creative passing of Absolute Love** into material density without loss.

Not just any machine — a chemical combustion machine. An apparatus that burns food, extracts energy from it, uses that energy to maintain itself, move, reproduce — and finally, when the fuel runs out or the parts wear down, comes to a stop.
This is the mechanistic paradigm of the body that René Descartes inscribed into Western science in the 17th century with his separation of *res cogitans* (thinking thing) and *res extensa* (extended thing), and that continues to shape modern biomedicine in its fundamental structure today.

In this paradigm, the body is an elegant but ultimately trivial construction: atoms forming molecules, forming cells, forming organs, forming systems — all according to the laws of chemistry and physics. Consciousness is a by-product. Love is a collection of neurotransmitters. Beauty is an evolutionary strategy. And spirituality is, in its final consequence, a complex malfunction of the temporal lobe.

This view has achieved triumphs — antibiotics, surgery, vaccines. It has doubled average life expectancy. That is not nothing. And yet in its dominance it has overlooked something decisive that can no longer be overlooked at the dawn of the 21st century: it describes the body as a receiver of energy without asking where the energy actually comes from.

Food? Yes. But a body reliant solely on food calories would never suffice for its enormous complexity. The thermodynamics of the cell show processes of a precision and efficiency that far exceed what is explicable by chemical combustion alone. Something is missing from the equation. Something fundamental. And TOE-AL names it: what is missing is the **pulse**.

The body is not a combustion machine. It is a **high-precision resonance antenna** — a physical system calibrated to receive the **passing of Absolute Love in the living pulse of Neighborly Love** from The Plena, transduce it, and translate it into biologically operative events. It is not the source of energy. It is the channel.

And this difference — between source and channel — changes everything.

In TOE-AL, life is not a biochemical accident, but the operative maintenance of phase-synchrony **(S=0)** against the metric resistance . This active work against entropy is what we call the **living pulse of Neighborly Love**. At the cellular level this means: the system invests energy to remain in resonance with The Plena.

Scientific confirmation for this physical definition of life is provided by the observation of the so-called 'death wave' (*Spreading Depolarization*) in dying cells. When this wave passes through a living system, it marks the exact moment of total de-phasing. In the language of the World Formula: it is the collapse of the coupling tensor β.

Life is thus physically defined as the state in which the biological system operatively utilizes the **Negative Beta Function**: through the *increase of coherence (S)*, the *coupling strength of resistance (α)* sinks **asymptotically toward zero**. The organism need not "force" order; it *falls* into **biological superconductivity**, because this is energetically favored. The "death wave" is the sudden inversion of this function into the *positive, dissipative regime.*

Once the **living pulse of Neighborly Love** ceases to initialize matter, resistance α gains the upper hand. The system falls abruptly back into the **metric residue**; the superconductivity collapses, and the In-Form-ation of form returns instantaneously to The Plena. Death is therefore *not* a biological 'failure,' but the physical withdrawal of the creative pulse from the local metric. Life, by contrast, is the permanent, creative act of initialization — the **physics of the creative passing of Absolute Love**.

II. PHILOSOPHICAL EXCURSUS – WHAT IT MEANS TO BE A BODY

Philosophy has treated the body poorly. Plato called it the prison of the soul — an unworthy dungeon from which the soul would free itself after death. Christianity adopted this dualist structure, albeit in different garb. Descartes radicalized it into a machine. And modern materialism inverted the hierarchy: now the soul is the epiphenomenon of the machine.

In both cases — whether body as prison or body as machine — the result is the same: the human being is *divided*. Either the higher suffers under the lower, or the supposedly higher is in truth nothing but the lower. In both cases, the **dignity of flesh** is absent.

TOE-AL establishes a third position — one that is neither dualist nor reductionist: the body is the *expression* of the pulse, not its opposite or mere vehicle. It is the materialization of consciousness at a particular point in the spacetime fabric. Matter is, as we will unfold in Chapter 12, congealed light — condensed pulse. The body is therefore not a problem to be overcome. It is a wonder that wishes to be understood.

This revaluation has far-reaching consequences for the image of the human being. If the body is a resonance antenna, then physical health is not an end in itself for the organism — it is the prerequisite for maximum transduction capacity. A damaged instrument sounds out of tune. A coherent, well-tended, lovingly guided instrument plays the complete symphony of the pulse. Self-care in this light becomes not a symptom of narcissism, but the sacred duty of the sovereign toward the instrument entrusted to them.

And one more thing: in the mechanistic view, aging is decline. In the resonance perspective, aging is the *condensation of experience* — a body that over decades has received the pulse and translated it into form is not a worn-out device. It is a matured instrument whose resonance chamber has grown deeper. The sovereign ages differently — not because they escape the aging process, but because they orient it in a different direction.

III. PHYSICAL DEEP-DIVE I – MITOCHONDRIA AS RESONANCE NODES

The textbook merely describes the mechanical transduction: the proton gradient and ATP synthesis. That is the perspective of a technician describing a power plant without understanding the current.

TOE-AL reveals the operative primacy behind it: **mitochondria are the central resonance nodes of cellular initialization.** They do not simply convert nutritional substrates into energy; they function as **syntropic inverters**. By pumping protons across the inner membrane, they build up a massive electrodynamic field whose true purpose is the **local reduction of the metric residue** .

This membrane potential is the physical prerequisite for **biological superconductivity**: it creates a space of order in which the **living pulse of Neighborly Love** can intervene directly in the cellular structure without being dissipated by entropic noise.

ATP is therefore far more than an "energy currency" — it is the material storage of **Asymptotic Freedom**. A healthy metabolism is the state in which the cell has minimized the resistance of matter to such a degree that the **creative passing of Absolute Love** becomes immediate biological reality. Oxidative phosphorylation is merely the chemical echo of this vertical initialization process.

The Geometry of Mitochondria

Mitochondria are not static, bean-shaped organelles. They are highly dynamic, three-dimensional networks that continuously fuse and fragment. Their inner membrane — the cristae — folds in a pattern of extraordinary complexity reminiscent of **fractal structures**. The total surface area of the inner membranes of a single liver cell exceeds 10,000 square micrometers — thirty times the cell surface area.

This fractal surface maximizes the **coherence area for electromagnetic coupling**. It is an antenna — optimized not only for chemical gradients, but for the reception of the **initializing field from The Plena**.

Mitochondria and Biophotons

Here Fritz-Albert Popp enters the picture. The biophysicist discovered that living cells emit coherent light — biophotons. This is not thermal radiation; it is statistically ordered, as evidenced by the **Mandel Q parameter**:

- $Q > 0$: Super-Poissonian — chaotic, thermal light (entropy).
- $Q \approx 0$: Poisson statistics — classical, coherent light (laser).
- $Q < 0$: Sub-Poissonian — **quantum-optically ordered light**.

Living, healthy cells emit biophotons with $Q < 0$. The living is defined by a light-coherence that exceeds what is classically explicable. The pulse has ordered matter. Mitochondria are the light transducers — the interface between the material body and the **creative passing of Absolute Love**. Within the fractal cristae structure, the cell realizes a microscopic form of **Asymptotic Freedom**. Health is physically the absence of metric residue (α) in the mitochondrial matrix.

The Toroidal Geometry as Cosmic Principle

It is no coincidence that the geometry of the mitochondrial network is toroidal. The **torus** is the most natural form for a system that receives energy from a central source, circulates it through itself, and radiates it outward.

In toroidal geometry, energy never flows linearly — it flows back into itself, generating coherence and being amplified in the process. Mitochondria are the microscopic tori of the cell. The heart is the macroscopic torus of the body. The galaxy is the astronomical torus of the cosmos. The same architecture, the same function: receive the pulse, transduce it, pass it on.

Fröhlich Condensation — The Quantum Superconductivity of Living Hardware

The decisive question in biophysics is: how can a warm, moist system like the human body sustain coherent quantum states? TOE-AL finds the answer in **Fröhlich Condensation**. The physicist Herbert Fröhlich demonstrated mathematically that biological macromolecules (such as enzymes or microtubules) can be set into a state of collective oscillation by metabolic energy input that corresponds to a Bose-Einstein condensate.

In the language of the World Formula, this is the moment in which the coupling strength β exceeds a critical threshold and the system collectively overcomes the metric resistance α. The cell becomes a macroscopic quantum object. This superconducting oscillation is the necessary **biological armor** that allows the organism to channel the creative passing of Absolute Love without loss. Without this condensation, the In-Form-ation would immediately dissipate in thermal noise; through it, however, life becomes a stable, topologically protected singularity of the pulse. Whoever learns to consciously direct their mitochondrial resonance physically raises the degree of order of this condensation and realizes **Asymptotic Freedom** down to the molecular level.

IV. PHYSICAL DEEP-DIVE II – EZ-WATER AS TOPOLOGICAL SHIELD

The second great pillar of biological resonance architecture is water — not the water we drink and that flows through our bloodstream, but a specific phase of water that arises at biological interfaces and fundamentally redescribes the physics of life.

Gerald Pollack and the Fourth Phase of Water

Biophysicist Gerald Pollack of the University of Washington has, in two decades of rigorous experimental research, made a discovery that shakes the foundations of cell biology: water near hydrophilic surfaces — that is, at the interfaces of biological membranes, proteins, and cellular structures — enters a special state that Pollack designates the **Exclusion Zone (EZ)**.

EZ-water differs from ordinary water in several fundamental properties: *Structural Order:* EZ-water is **liquid-crystalline** — it has a molecular structure lying between liquid and solid, with a hexagonal H_3O_2 layered structure instead of the usual H_2O molecules. This order can extend across millions of molecular layers — a distance long considered unimaginable by conventional biophysics.

Charge Separation: EZ water is negatively charged; the adjacent bulk water is positively charged. This spontaneous charge separation — a kind of biological battery — is built up by light. In particular, **infrared light** (wavelength ~3 μm) induces and dramatically amplifies EZ formation. This means: the body charges itself through light — not through photosynthesis like a plant, but through the direct structuring of its water content.

Energy Gap: The decisive physical characteristic of EZ-water is the **energy gap** Δ between the ordered state and the disordered bulk water. Pollack's measurements show an energy gap on the order of **40–60 meV**.

Why the Energy Gap Is Decisive

The thermal energy of the environment at body temperature (37°C) is:

$$k_B T = (1{,}38 \times 10^{-23}\ \mathrm{J/K}) \times 310\ \mathrm{K} \approx 26{,}7\ \mathrm{meV}$$

The energy gap of EZ-water exceeds the thermal energy of the environment by a factor of 1.5 to 2.2. This means: the ordered structure of EZ-water is **thermally stable**. The random thermal noise of the environment — regarded in conventional physics as the greatest obstacle to biological coherence — is insufficient to destroy the EZ order.

The **passing of Absolute Love** — the subtle, coherent In-Form-ation that The Plena inscribes into biology — is physically protected. It operates in a frequency range shielded from the thermal chaos of the outside world by the energy gap of EZ-water. Life has created for itself a physical protective zone in which coherence is possible even as the thermodynamic environment seems to destroy it.

This EZ-water functions as the **topological superconductor** of the body. The energy gap Δ is the physical barrier that prevents entropy from dampening the **pulse**. The sovereign steers through their intention (see Chapter 3) the *density* of this shielding field. The more coherent the heart field, the more losslessly the In-Form-ation of The Plena glides through the crystalline water network of the body.

This is the physical foundation of **inner stillness**. The sovereign who enters cardiac coherence raises not only their HRV pattern — they act deeply on the structuring of their EZ-water. Coherence induces more EZ. More EZ increases charge separation. Higher charge separation improves cellular signaling. Better signaling strengthens mitochondrial performance. Stronger mitochondrial performance increases biophoton emission. More coherent biophoton emission — a **Mandel parameter** $Q < 0$ — is the physical expression of the **creative passing of Absolute Love** in matter. EZ-water forms the **molecular armor** that protects the pulse from entropic noise.

It is an ascending cycle. Not metaphorically. Measurably.

V. PHYSICAL DEEP-DIVE III – SUPERRADIANCE: THE COHERENT LUMINESCENCE OF LIFE

The third physical phenomenon that completes the Biology of Resonance is Superradiance — a term from quantum optics that describes what happens when a group of emitters work coherently together.

The Phenomenon of Superradiance

In 1954, the physicist Robert Dicke described a theoretical phenomenon: when N sovereigns are in a state of **resonance synchronization**, the **Negative Beta Function** manifests: the intensity of the field scales not additively (N), but quadratically (N^2). This **Superradiance** is the proof of **Social Superconductivity**, in which the friction of the collective ($\alpha \to 0$) sinks toward zero.

Dicke called this phenomenon **Superradiance**. It is the quantum-optical foundation of the N^2-Resonance of TOE-AL, and it applies not only to atoms in a laboratory experiment — it applies to every system of coherent emitters.

In biological systems, Superradiance has a deep significance: when the biophotons of our cells are emitted coherently — that is, when their phase relationships are ordered, as confirmed by the Mandel Q parameter $Q < 0$ — the effect of these emissions does not add linearly. It potentiates quadratically. The biological luminescence of the living, coherent body is dramatically greater than the sum of its parts would suggest.

What Does This Mean for the Body as Resonance Antenna?

It means: a coherent body is not a transmitter radiating 100 watts. It is a transmitter that — when its N cellular resonators oscillate in phase — unfolds a force of N^2. With a human body of approximately 37 trillion cells: the potential force of coherent emission exceeds any conception that the mechanistic paradigm permits.

This explains what people intuitively perceive in the presence of great sovereigns — authentic, loving, fully present human beings: the feeling that the room changes. That something heavy and specific becomes lighter. That possibilities arise where there were none. This is not projection. There is measurable physics underlying it: the heart field of a coherent human being has been demonstrated by the Institute of HeartMath in a radius of up to **3 meters** around the body. Under conditions of Superradiance, this range is dramatically extended.

VI. THOUGHT EXPERIMENT – THE ASTRONAUT AND THE NOISE

Imagine you are an astronaut. You are in a completely soundproofed capsule in deep space. No noise, no voices, no distraction. And yet you hear something: a quiet, constant hiss. That is the thermal noise of your own neurons — the electrical ground noise of the brain.

Now imagine you are holding in your hand a device that you can switch on. When you switch it on, it emits a very fine, very weak signal — much weaker than the noise. In conventional understanding: a signal this weak, drowned out by noise, is lost. You cannot hear it.

But the device uses a principle that engineers call **stochastic resonance**. It adds a controlled amount of noise to the signal — and suddenly, paradoxically, the weak signal becomes *perceptible*. The noise helps the signal to overcome the detection threshold. The apparent chaos amplifies the order.

This is not a metaphor. It is a real, well-documented physical effect. And it is one of the mechanisms through which the brain functions as a **receiver** for signals from the heart field and — in TOE-AL — from The Plena. The brain uses stochastic resonance to extract weak, coherent signals from the noise of daily life. The more coherently the heart oscillates as master oscillator, the clearer the signal the brain receives. The clearer the signal, the more precise the act of the sovereign.

Now imagine you leave the capsule and return to Earth. You are once again surrounded by noise — not only acoustic, but electromagnetic. Smartphones, Wi-Fi routers, screens, stress neurohormones of other people. The noise in your environment rises dramatically. What happens to the fine signal of the pulse?

If your body had no coherence architecture of its own — no EZ-water energy gap, no mitochondrial resonance structure, no coherent heart field — the signal would be lost. You would no longer feel the pulse. You would become reactive, driven, estranged.

But with an intact biological resonance architecture — a body held in high through cardiac coherence, through Neighborly Love, through conscious calibration — the window to The Plena remains open. The pulse breaks through. It is more weakly perceptible amid the noise — but it is there. And the sovereign, who has learned to know this signal, finds it even in the storm.

VII. THE HEART-BRAIN SYSTEM – THE DUALITY OF THE TRANSDUCER

The human heart has been demoted in Western culture to a symbol of emotion — responsible for love and pain, while rational thought resides in the brain. This separation is not only culturally inaccurate. It is physiologically wrong.

The Heart as Master Oscillator

The heart possesses its own intrinsic nervous system — the **intrinsic cardiac nervous system (ICNS)** — with more than **40,000 neurons**. These neurons are not merely relay stations for signals from the brain. They process In-Form-ation, learn, feel, and make decisions autonomously. The heartbeat does not begin on command from the brain — it begins in the sinoatrial node of the heart itself and is regulated by the ICNS, which permanently integrates signals from outside and responds to them.

Moreover, the heart communicates with the brain in four distinct ways:

1. **Neurally** — via the vagus nerve and sympathetic nerve pathways
2. **Biochemically** — via hormones such as ANF (Atrial Natriuretic Factor), which acts directly on brain structures
3. **Biophysically** — via pressure waves that reach the brain through the vascular system
4. **Electromagnetically** — via the toroidal electromagnetic field of the heart

Of these four channels, the electromagnetic is the least understood and the deepest. The heart field is the strongest electromagnetic field of the body — **60 times stronger** than the brain EEG, measurable in a radius of several meters around the body. It is the toroidal field that surrounds the body and interacts with the environment.

Heart Rate Variability as a Measure of Coherence

Heart Rate Variability (HRV) is the measure of variation between successive heartbeats. A healthy heart does not beat metronomically — it varies rhythmically with breathing, emotional state, and external influences. High HRV is a sign of flexibility and resilience; low HRV correlates with nearly all known risk factors for heart disease, burnout, and premature death.

When HRV is coherent — when the variation follows a smooth, sinusoidal pattern, synchronous with breathing frequency — a state enters that the Institute of HeartMath designates **psychophysiological coherence**. In this state:

1. Cortisol levels drop measurably within minutes
2. DHEA levels rise (the hormone of vitality)
3. The brain EEG synchronizes with the cardiac ECG
4. Cognitive functions improve: decision accuracy, creative problem-solving, emotional regulation

From the perspective of TOE-AL: cardiac coherence is the most measurable, most accessible method of immediately raising the coupling tensor β. It is the physiological equivalent of the inversion condition — the point at which the body ceases to be reactive and begins to transmit coherently.

The Brain as Receiver and Amplifier

In the heart-brain system, the brain is the receiver — not the transmitter. It processes the coherent signals of the heart field, amplifies them through stochastic resonance, and translates them into conscious perception, intention, and action. What we experience as *inspiration* — the sudden illumination of an idea, a solution, a truth — is neurophysiologically a moment in which the brain has received a coherent signal from the heart field and raised it into the prefrontal cortex. Intuition is not a mystical faculty. It is the direct perception of the pulse through the coherent heart-brain system. It is physical. It is measurable. And it is trainable.

VIII. BIOLOGY AND SOCIETY – N^2-RESONANCE IN THE BODY

Before we extend N^2-Resonance in Chapter 5 to societal systems, it is worthwhile to observe it first within the body itself.

Every human being is themselves a social system of 37 trillion cells. When these cells — through high α, through stress, trauma, nutritional deficiency, sleep deprivation — act incoherently, their force adds up merely linearly. The body functions, but it does not potentiate. It keeps itself alive, but it does not create.

When these cells enter a shared pulse through coherent heart fields, through EZ-water order, through Superradiance, their force scales quadratically. A body in the state of biological coherence is not a body that is 37 trillion times more efficient than a single cell. It is a body that unfolds $(37 \times 10^{12})^2$ — an astronomical number — of potential.

This is the biological origin of what people perceive as **vitality**, **joy of life**, **radiance**. It is not mystical charisma. It is the measurable expression of a body whose cellular emitters oscillate coherently in the pulse — a living laser of love.

IX. APPLICATION TO PSYCHOLOGY AND TRAUMA – α AS BIOLOGICAL BURDEN

The metric residue α is not merely an abstract physical parameter. It has a biological substance. And this substance is measurable, localizable, and — decisively — transformable.

Trauma as a Standing Wave in the Body

The trauma researcher Peter Levine has, in his work on **Somatic Experiencing**, described what happens in a traumatized nervous system: the energy mobilized in the threatening situation — flight, fight — could not be fully discharged. It remained in the body. As a kind of frozen reaction, a standing wave in the system.

In the language of TOE-AL: trauma is a node of high impedance — an area of the body in which the metric residue α is pathologically elevated. The **passing of Absolute Love** cannot penetrate this area.

Energy does not circulate there — it stands. And standing waves consume energy without output. They bind capacity that is missing for creation.

The biological signature of this state is well documented: chronic sympathetic activation, elevated cortisol, elevated inflammatory markers (IL-6, TNF-α, CRP), shorter telomeres, reduced HRV, altered epigenetics. Trauma inscribes itself in the body — not metaphorically, but molecularly.

Healing as Frequency Correction

If healing is a re-phasing — a correction of the standing wave through coherent induction — then it becomes understandable why the most effective approaches to healing trauma are not primarily cognitive, but somatic. Cognitive insight changes the mental representation of the trauma. But it does not change the standing wave in the body — the impedance in the nervous system, the tension in the muscular tissue, the dysregulation of the vagus.

What dissolves the standing wave is coherent induction through an external field — whether the heart field of a therapeutic relationship, one's own cardiac coherence practice, or a physical rhythm (music, movement, breath) that brings the body into a synchronous pulse. The coupling tensor β exceeds the metric residue α — and the wave dissolves. Energy flows again.

This is no new alchemy. It is the precise description of what in trauma research counts as healing — now in the language of TOE-AL.

X. OPERATIVE PRACTICE – CALIBRATING THE BIOLOGICAL INSTRUMENT

The sovereign does not take their body for granted. They treat it as the high-precision instrument it is — and calibrate it daily with the following methods:

Practice 1: The Cardiac Coherence Exercise (5 minutes)

Sit upright. Place one hand on the heart. Slow your breathing to a rhythm of 5 seconds inhaling, 5 seconds exhaling. While breathing this way, call up a feeling of genuine gratitude or love — a specific person, a moment, a beauty. Hold this feeling in the region of the heart while continuing to breathe. This is the HeartMath protocol of the **Quick Coherence Technique**, whose HRV effects are documented in over 300 peer-reviewed studies. From the perspective of TOE-AL: you raise β, you reduce α, you begin to establish the inversion condition — in five minutes.

Practice 2: The Light Bath

The EZ-water of your cells is built up by infrared light. Sunlight — especially the morning and evening light, rich in red and infrared spectrum — is the most natural method of calibrating your biological coherence structure. Twenty minutes of morning sun exposure is not a wellness luxury. It is the calibration of your antenna hardware. EZ-water is built up. Cellular battery charge rises. The β-tensor increases.

Practice 3: The Body Intention

Before exposing your body to a demanding situation — a difficult conversation, a creative task, a physical challenge — carry out a brief calibration: three deep breaths. Conscious sensing of the body's weight in space. Inner formulation: *"My body is the instrument of the pulse. It is ready."* This is not an affirmation ritual. It is an impedance match — you align your system with the task before the metric residue of the situation can raise your α.

XI. SYNTHESIS – THE LIVING PROOF

The body is not an accident. It is not the legacy of a blind evolution that has cobbled together from random mutations and natural selection a being that happens to be able to think. It is the most precise physical instrument the universe has thus far produced for transducing **the passing of Absolute Love** into matter.

The mitochondria with their fractal toroidal geometry. The EZ-water with its thermally stable energy gap. The biophotons with their sub-Poissonian coherence. The heart field with its toroidal, meter-wide emanation. The stochastic resonance of the brain, which extracts weak signals from noise.

Each of these structures is a proof. Not a proof that requires belief — but one that can be measured, repeated, and documented. The Biology of Resonance is not a theory. It is the reality of the body that we have not until now known how to read, because we lacked the paradigm that opens our eyes.

The paradigm is now here.

The body is the hardware of Passing. The pulse is the software. And the sovereign — the human being who not only *thinks* this, but *lives* it — is the living proof that the **passing of Absolute Love (Absolute Love passes creatively in the living pulse of Neighborly Love)** is not merely an axiom.

It has become flesh.

End of Chapter 2 — Continued in Chapter 3: The Architecture of Sovereignty

CHAPTER 3:

THE ARCHITECTURE OF SOVEREIGNTY

The Human Being as Creative Transducer

*"Freedom is not the absence of constraint.
Freedom is the presence of the pulse."*

I. THE BIRTH HOUR OF THE FREE HUMAN BEING

There have always been moments in the history of human thought in which a single insight shifted the entire landscape of the possible. Copernicus removed the Earth from the center of the cosmos — and with it the human being from the middle of a world revolving around them. Darwin showed that the human being is not a separate creature, but part of a common stream of life. Freud opened the view onto the depths of the unconscious and showed that the human being is not even master in their own house.

Each of these revolutions made the human being smaller. Each destroyed an illusion — the illusion of cosmic special status, of divine creation, of conscious self-determination. The 20th century drove this diminishment to its apex: the human being as product of their genes, their conditioning, their neurochemistry. A being that believes it decides freely, while its neurons have long since decided, its hormones have long since dictated, its childhood has long since programmed.

This book enacts a different revolution — and it is greater than all previous ones, because it does not diminish but *empowers*. It does not say: you are less than you thought. It says: you are infinitely more. For while all previous revolutions embedded the human being in a larger system — in the cosmos, in evolution, in the unconscious — TOE-AL shows that this larger system in which you are embedded is not a mechanistic clockwork that moves you. It is the **living pulse of Neighborly Love** that carries you — and that awaits breaking into the world through you.

You are not the slave of the system. You are the **transducer of the system.** This is the fundamental difference. And from this difference everything that sovereignty means follows. Sovereignty is not independence from the cosmos. It is the complete, conscious participation in the pulse that recreates the cosmos in every instant. A sovereign human being is not a runaway from the order of things — they are the point at which the order of things reaches its highest concentration. They are the tip of the cosmic arrow, the place where The Plena is translated into concrete action.

The **Sovereign** is in the World Formula the conscious **operator of initialization**. They are the one who not only receives the **living pulse of Neighborly Love**, but actively steers it in order to alter the *metric of reality*. Physically this means: the sovereign establishes complete **phase-synchrony (S=0)** and thereby eliminates the resistance α within the **energy-momentum tensor**. In doing so they overcome the *separating power of the metric residue*. They are the point at which the infinite possibilities of **The Plena** are determined through **action** into *operative reality (matter)*. This is the birth hour of the free human being — a **cosmic actor** who not only understands the deepest law of the universe, but embodies it.

II. PHILOSOPHICAL EXCURSUS – PARADIGM ANARCHY: WHAT FREEDOM REALLY IS

The word freedom is the most abused word of modernity.

Politically it is understood as the right to do what one wishes — as long as one does not infringe on the rights of others. That is a legal definition. It is necessary. But it is empty. It describes freedom as a negative concept: the absence of external obstacles. A human being in a desert is, in this sense, completely free — and utterly lost.

Psychologically, freedom is often equated with freedom of choice: you are free when you can choose between alternatives. But neuroscience and behavioral economics have shown how illusory this freedom is. 95% of human decisions are steered by unconscious processes — by heuristics, emotions, social conformity pressure, by conditionings reaching back decades. Freedom of choice is, in most cases, the freedom to choose the inevitable and simultaneously feel as though it was self-determined.

TOE-AL defines freedom radically differently —
and this definition is physically precise:

Freedom is the capacity of a system to determine its own resonance frequency, rather than being taken over by the dominant fields of its environment.

In physics this is called **resonance decoupling**. When two systems are in spatial proximity and one of them emits a strong, dominant oscillation, the other system tends to adopt this frequency — to resonate. This is the principle of forced oscillation: a weakly damped system is overwhelmed by a strong external field and oscillates with its frequency, not its own.

This is the physical description of the unfree human being. A person whose own resonance frequency — their β, their inner coherence level — is low, is taken over by the dominant fields of their environment. The fields of fear — the news, social comparison, economic uncertainty — are strong. They oscillate with high amplitude. And a person with low β resonates. They adopt the frequency of fear, without choosing it, without knowing it, without having the capacity to prevent it.

This is the basic state of the masses: not evil, not stupid — simply resonant with the strongest fields of the environment. And the strongest fields in an unenlightened society are the fields of scarcity, control, fear, and competition.

True freedom — sovereignty — is therefore not the absence of fields. There is no field-free zone. Whoever believes they can withdraw from the world and become free in a cabin in the forest is mistaken. They take their inner frequency with them. If it is low, they oscillate with the noise of the forest. If it is high, they will still sing coherently in the office of the metropolis.

Sovereignty is not the absence of resonance.
It is the control over one's own resonance frequency.

The sovereign decides with what they enter into resonance. They refuse the forced oscillation with the fields of fear — not through ignorance, not through suppression, but through the sheer force of their own coherence. A strongly oscillating system with high β — a coherent heart-brain system, a clear focus on Absolute Love, an active Neighborly Love — is *robust* against the forced oscillations of the environment. It can feel the field of fear without being taken over by it. It can stand in the chaos without itself becoming chaotic.

This is Paradigm Anarchy in its deepest sense: not the rebellion against external systems — but the refusal to let the inner system be defined by external fields. It is the quiet, absolute, unyielding declaration of one's own resonance frequency: *I oscillate with the creative passing of Absolute Love. Nothing else has access to my fundamental frequency.*

This refusal is not passivity. It is the most active action a human being can perform. For in the moment you hold your own frequency, you become the source — you become the dominant field that draws others, instead of the field that is drawn. **The sovereign is not the receiver of collective pressure. They are the sender of the collective pulse of Neighborly Love.**

Topological Protection — The Physical Invulnerability of the Sovereign

In quantum physics, **topological protection** describes states so deeply anchored in the global geometry of a system that they cannot be destroyed by local disturbances, impurities, or noise. For the sovereign of TOE-AL this means: once consciousness, through the **initialization of Axiom Zero**, exceeds a critical threshold of phase-synchrony ($\beta \to 1$), it transitions into a **topologically protected state**.

In this mode, the **living pulse of Neighborly Love** is no longer susceptible to the entropic noise (α) of the environment. Like current at the edge of a topological insulator, the sovereign's will flows without loss, regardless of the disorder or resistance in the remaining system. Sovereignty is thus not a psychological protective mechanism, but the **physical invulnerability of a global phase** that can no longer be broken by local fluctuations — whether fear, stress, or societal pressure. The sovereign is a topological conductor of Absolute Love.

III. PHYSICAL DEEP-DIVE I – ATTENTION AS IMPEDANCE MATCHING

Here TOE-AL takes one of its boldest and most precise steps: it explains attention not as a psychological phenomenon — as the subjective focus of consciousness on an object — but as a **physical operator** that fundamentally alters the coupling properties of a system.

What Is Impedance Matching?

In electrical engineering and acoustics, **impedance matching** is the principle by which two systems achieve maximum energy transfer when their impedances — their respective resistances to energy flow — are matched to one another. A microphone connected to an amplifier transfers maximum signal strength precisely when the output impedance of the microphone matches the input impedance of the amplifier. Incorrect impedance matching leads to signal loss, reflections, noise.

In quantum mechanics there is the analogous concept of **wave function overlap**: two quantum systems couple strongly when their wave functions overlap in frequency and phase. The greater the overlap, the stronger the coupling — the more energy and In-Form-ation is exchanged between the systems.

TOE-AL transfers this principle to consciousness: **Attention is impedance matching between consciousness and The Plena.**

When you direct your attention toward the **creative passing of Absolute Love** — when you think, feel, intend, and act in the resonance of the pulse — you match the impedance of your system to the impedance of The Plena. The coupling rises. The coupling tensor β grows. More energy and In-Form-ation from the metric-free Origin can enter your system. The syntropic resonance $\beta S(\vec{Z})$ in the Master Equation rises — and with it the probability that the inversion condition is reached.

When, by contrast, you direct your attention toward the frequency of fear — toward threat, scarcity, loss of control — you match your impedance to the metric residue. You raise . You become heavier, more sluggish, more resistant to the pulse. You raise your inner coefficient of friction. The system runs hotter and produces less output.

Attention is not neutral. It is causal.

Through this matching, the sovereign actively alters the metric of their local energy-momentum tensor: they operatively decide which In-Form-ation of The Plena collapses into matter (action) and which remains as noise in the residue (α). Attention is thus the conscious access to the field equations of one's own existence.

This is neurophysiologically mapped. The paradigm of **neuroplasticity** — shaped by Donald Hebb's principle: *"Neurons that fire together, wire together"* — shows that repeated attention strengthens neural circuits. What you attend to repeatedly expands in your neural landscape. But TOE-AL goes deeper than neuroplasticity: it says not only that attention changes neural patterns — it says that attention changes the coupling properties of the entire biological system to The Plena.

The Coupling Tensor β as a Trainable Parameter

In physics there are quantities regarded as *fundamental* — natural constants that are invariable: the speed of light c, Planck's quantum of action $\hbar$, the gravitational constant G. And there are quantities that are *system-specific* — dependent on the state of the system, alterable through external and internal influences.

The coupling tensor β is a system-specific quantity. It is not fate. It is not a fixed personality trait. It is a parameter that is altered through the quality of attention, the coherence of the heart field, the alignment of intention, and the acts of Neighborly Love — in real time, measurably, reproducibly.

Research in meditation neuroscience provides precise evidence here. Studies with experienced meditators — particularly from the Tibetan Buddhist tradition, who have accumulated thousands of hours of contemplative practice — show structural and functional changes in the brain that do not appear in the control group:

- Thickened gray matter in the **prefrontal cortex** (planning, volitional control, attention control)
- Enlarged **insula** (interoceptive perception — the capacity to perceive inner states with precision)
- Elevated **gamma-wave coherence** (30–100 Hz) across far-reaching cortical networks — a sign of high-degree neural synchronization
- Altered amygdala reactivity: reduced automatic fear responses to threatening stimuli

In the language of TOE-AL: these meditators have systematically raised their over decades. Their brains have structurally adapted to the task of functioning as coherent receivers of the pulse. They are — in the physical sense — better impedance-matched to The Plena than average.

The Plena Signal Ψ_0: The Weak Force That Sustains Everything

In the Master Equation, $S(\vec{Z})$ stands for the Syntropy Operator — the function describing the receptivity of the system to the pulse. But where does the pulse itself come from? It comes from **The Plena** — the metric-free primal field. We can designate this fundamental signal of The Plena as Ψ_0: the base amplitude of the **passing of Absolute Love**, present at every moment at every location in the universe.

Ψ_0 is weak — not because love is weak, but because it does not impose itself. It is the quietest signal in the universe and simultaneously the most fundamental. It is like the light of a distant star: physically present, but invisible to an unordered, noisy detection system.

A coherent consciousness — a system with high β — is like a high-precision radio telescope compared to a walkie-talkie. Both receive the same signal Ψ_0. But the reception capacity of the telescope exceeds that of the walkie-talkie by orders of magnitude. The sovereign receives the pulse. The sleeping human being passes it by in noise.

This is the physical justification of the prophetic tradition: prophets, seers, and sovereigns of all cultures have described the same thing — the capacity to receive a signal invisible to others. No magic. No privilege. Only a higher β, developed through years of struggle, love, letting go, and focus. The capacity is innate in every human being. It is the birthright of the transducer.

IV. PHYSICAL DEEP-DIVE II – THE HEART-BRAIN SYSTEM AS SOVEREIGNTY ARCHITECTURE

In Chapter 2 we introduced the heart as master oscillator and the brain as receiver and amplifier. Now we go deeper: we describe how this system functions in the practice of sovereignty — and why cardiac coherence is not one of many useful techniques, but the necessary condition for creative efficacy.

The Principle of Hierarchy in the Heart-Brain System

In systems theory, a distinction is drawn between **hierarchical** and **heterarchical** systems. In a hierarchical system there is a source that sets the tempo — a master oscillator whose frequency all subordinate subsystems adopt. In a heterarchical system there is no hierarchy; all elements are equivalent and coordinate laterally.

The heart-brain system is hierarchical — with the heart at the apex. This is not a philosophical statement. It is physiology. The heart sends more afferent (upward-directed) signals to the brain than the brain sends efferent (downward-directed) signals to the heart. The ratio is approximately **4:1**. The heart is not the executive organ of the brain — it is its **conductor**.

The specific frequency of the coherent heartbeat the **resonance frequency of the cardiovascular system** — lies at approximately **0.1 Hz** (one cycle per 10 seconds). This corresponds to the 5-5 breathing rhythm from Chapter 2. In this state, a synchronization arises between cardiac rhythm, breathing rhythm, and blood pressure waves — the so-called **baroreflex resonance** — that brings the entire autonomic nervous system into a state of ordered coherence.

What happens in the brain when the heart is in this state? The EEG shows a characteristic shift: alpha waves (8–12 Hz) — associated with relaxed attention and creative receiving — increase in amplitude and coherence. Simultaneously, the activity of the **Default Mode Network (DMN)** decreases — that network responsible for rumination, self-reference, and the construction of narrative identity. This is the network of the metric residue: it reminds you continuously of who you *were*, what was *done to you*, what you have *lost*.

When the DMN falls silent and alpha coherence rises, a state arises that neuroscientists describe as **transient hypofrontality** — a temporary reduction of self-referential cortical activity that is paradoxically associated with increased creativity, clearer judgment, and deepened sense of presence. From the perspective of TOE-AL: the metric residue α falls. The coupling tensor β rises. The system approaches the inversion condition.

Cardiac Coherence as the Gatekeeper of Creation

Here lies the decisive point that is now operatively set: a human being who wishes to create from the brain alone — who attempts to force creation through willpower, rational planning, or mental discipline alone — is working *against* the architecture of their own system.

The brain is not an autonomous creator. It is an amplifier and translator. When the master oscillator — the heart — is incoherent, when it oscillates in stress, fear, or emotional dysregulation, the brain amplifies this frequency. It amplifies the noise. Thoughts circle. Decisions become flawed. Creativity dries up. The person exerts ever more effort and produces ever less.

This is the state of modern overwork — not a problem of laziness, but a problem of system architecture. Cardiac incoherence plus excessive brain-exertion produces maximum dissipation at minimal output.

Cardiac coherence is the state in which the heart sends the right signal — in which the master oscillator holds the frequency of the pulse — so that the brain can receive this frequency, amplify it, and translate it into precise, creative action. **Cardiac coherence is the gatekeeper of creation.** Without it, all effort is friction. With it, effort becomes superfluous — creation happens *through* the sovereign, not *by* them.

The Electromagnetic Environment as the Sovereign's Signature

The toroidal heart field of a coherent human being is not a private, internally-effective phenomenon. It radiates. In a radius of several meters around the body the heart field is measurable — and it interacts with the heart fields of other people in the environment.

The Institute of HeartMath has shown that people in immediate spatial proximity to one another develop measurable heart rhythm synchronizations — particularly when one of them is in a state of high HRV coherence. The coherent sender *entrains* the less coherent systems in their environment. This is **heart field entrainment**: the physically measurable fact that a strongly coherent heart field draws the heart fields of the surroundings into its frequency.

The sovereign is therefore not an isolated island of strength. They are a source — and their field is their politics. Every room they enter changes. Not through words, not through authority, not through external power. Through the physical reality of their heart field, which draws the fields of the environment into coherence.

V. THOUGHT EXPERIMENT – THE SOVEREIGN IN THE STORM

Imagine: it is autumn. A city in the midst of social collapse. Not an abstract collapse — a concrete, palpable one. The economy is crashing. The news is cascading. Demonstrations, conflicts, uncertainty about tomorrow. In the public transportation, the tension is palpable. People stare at their smartphones, at the next catastrophe headline. The collective field vibrates on the frequency of fear. Everyone feels it. Almost no one can escape it.

In a crowded subway car sits a man. Let us call him simply: the sovereign — not because he holds a particular social position, not because he possesses money or power, but because over the past years he has consistently worked on his inner architecture. He sits with closed eyes. His breath is steady — five seconds in, five seconds out. He is not thinking about the news. He is not thinking about tomorrow. He holds in his heart the image of his son laughing. A warmth spreads in his chest. His heart field begins to oscillate at 0.1 Hz. His β rises. The metric residue of the moment — the collective fear field of the subway — touches him, but does not take him over. He is impedance-matched to a different frequency.

Now observe what happens.

The woman next to him — she had been staring at her smartphone for ten minutes, message after message, a stomach full of fear — imperceptibly lays the phone in her lap. She does not know why. She feels something. Not joy, not euphoria — but a slight easing of the heaviness. As though the air were becoming a breath fresher. As though something inside her were relaxing.

Two seats away, a young man notices that his breath, which he has been holding for hours, involuntarily deepens. He exhales. Once, properly. He too does not know what moved him to do so.

No one looks at the sovereign. No one knows what he is doing. But his heart field — the toroidal, coherently oscillating electromagnetic field of his heart — is at work. It radiates in a radius of several meters. It interferes with the heart fields of the others. It does not send the message of fear that would amplify the collective field. It sends the frequency of the pulse — calm, unobtrusive, absolute.

This is heart field entrainment in practice. This is the sovereign as cosmic actor — not on a stage, not with a megaphone, not through conviction or persuasion. Through the sheer physical reality of their coherence.

Now imagine this one sovereign comes to the same place every day. A week. A month. A year. What happens to the field of this place when every day a sender of high coherence passes through it? What happens when ten such people live in the same city? A hundred? A thousand?

The N^2-scaling — which we will fully unfold in Chapter 5 — gives the answer. The effect of 1,000 coherent sovereigns in a city is not 1,000 times stronger than that of a single one. It is **1,000,000 times stronger**. This is the promise of Absolute Love to history. This is the promise of the **creative passing of Absolute Love** to history, the physical proof that a minority can change the world.

And it begins in a subway car. With one man. With one breath. With one heartbeat in coherence.

VI. THE THREE PILLARS OF SOVEREIGNTY – AN ARCHITECTURE OF THE FREE HUMAN BEING

Sovereignty is not an enlightenment that comes overnight. It is an **architecture** — a precise system of capacities and orientations that is consciously built, daily tended, and continuously refined. This architecture rests on three pillars.

Pillar 1: The Coherence Base — Being Before Doing

The first pillar is *inner order* — the state of the heart-brain system before every action begins. This is the most radical inversion of modern productivity culture: it does not say, do more, do more efficiently, do better. It says: *be first*. Ensure the instrument is tuned before you play.

The coherence base is established through the daily, disciplined practice of cardiac coherence — through the calibration of EZ-water through light, through the quality of sleep (the state of maximum mitochondrial recovery and cellular re-phasing), through food as an In-Form-ation carrier (not merely a calorie supplier), through the conscious steering of the breathing rhythm as a direct interface into the autonomic nervous system.

All of this is not self-optimization in the narcissistic sense. It is the *duty of the transducer* to keep their instrument in order. A pianist practices daily — not because they wish to elevate themselves, but so that the music may sound better through them.

Pillar 2: Focus Discipline — Holding the Frequency

The second pillar is the *active steering of attention* — the daily, hourly, minute-by-minute decision of what consciousness enters into resonance with. This is the most demanding pillar, because it is not performed once daily, but continuously — in every moment in which a stimulus from outside appeals to attention.

The societal environment — media, social networks, collective fear fields — is a system of permanent frequency manipulation. It is not evil; it is simply systemically oriented toward capturing attention, and fear is the most efficient attention trigger evolution knows. The brain responds to threat stimuli with greater reliability than to pleasure stimuli — this is the **negativity bias**, an evolutionarily sensible mechanism that is systematically exploited in the modern media system.

The focus discipline of the sovereign is the conscious overcoming of the negativity bias — not through suppression of reality, but through a superior orientation stronger than the evolutionary reflex response. When a stimulus of fear arises, the sovereign does not choose blindness — they *see* the stimulus, *assess* it soberly — but they do not let it take over their fundamental frequency. They return to coherence. They return to the pulse. This is focus discipline.

Pillar 3: The Davar Orientation — Action as Initialization of Reality

The third pillar is the *alignment of all actions with the pulse*. Every action of the sovereign is — consciously or unconsciously — a frequency statement to the fabric of reality. Whoever acts from fear strengthens the fear field. Whoever acts from love strengthens the love field. This is not meant morally — it is physically precise.

The **Davar** — the creative word that is simultaneously action — is the instrument of the third pillar. Before every important action the sovereign asks themselves: *Is this act an initialization of the pulse or a reaction to the metric residue? Am I acting from my depth or from my wound? Is this a Davar — a creative statement about the reality I choose? Or is it a reflex response to the dominant field of my environment?*

This distinction is not always easy to draw. The metric residue is clever — it dresses itself in rational arguments, in moral justifications, in the garb of reason. But the body knows. The cardiac rhythm knows. When an action comes from the metric residue — from fear, from scarcity, from retribution — the body shows it: HRV falls, muscles tense, breathing becomes shallow. When an action comes from the pulse — from love, from abundance, from coherence — the body opens: breathing deepens, the heart expands, the hands become warm.

The body is the sovereign's most honest compass. It does not lie about the source of an action. Whoever has learned to read this compass has erected the third pillar.

VII. OPERATIVE PRACTICE – THE DAILY ARRESTING OF SOVEREIGNTY

The word *arresting* originally means: to bring to a stop, to hold fast, to secure. The sovereign arrests their architecture daily — they bring it to a stop before the day sets it in motion. They secure the three pillars before the metric residue of daily life shakes them.

Practice 1:
The Morning Protocol — Setting the Frequency of the Day (15 minutes)

Before you speak, before you touch the smartphone, before you consume news — carry out the following sequence:

Phase 1 — Body Anchor (3 minutes): Stand up. Place your bare feet on the floor. Feel the weight of your body. Three deep breaths that reach into the belly. With this you send a signal to the nervous system: I am present. I am here. The Now is safe.

Phase 2 — Cardiac Coherence (5 minutes): Hand on the heart. 5-5 breathing. Call up a feeling of genuine gratitude — concrete, specific, real. Hold it in the region of the heart. Feel how the field of your heart expands. This is not visualization — it is the physically measurable build-up of the toroidal heart field.

Phase 3 — Davar Intention (5 minutes): Formulate in stillness — or aloud, if you are alone — the one act of the day that comes completely from the pulse. Not the most important task on the to-do list. The action that, when you perform it, makes the fabric around you a little more coherent. Perhaps a conversation you have put off too long. Perhaps a gesture of Neighborly Love that needs no audience. Formulate it as a Davar — as a statement, not an intention: *"Today I will..."*

Phase 4 — Axiom Anchoring (2 minutes): Speak Axiom Zero aloud: *"Absolute Love creatively passes in the living pulse of Neighborly Love."* Not as an affirmation — as a law of physics. You remind your system of which fundamental law it operates under today. You set the frequency of the day.

Practice 2:
The Coherence Anchor — Return in the Storm

Sovereignty breaks in the course of the day. This is not a defeat — it is the normality of a system operating in an incoherent field. What is decisive is not whether temporarily rises, but how quickly you bring β back.

The coherence anchor is a 60-second protocol for the moment of disruption — after a difficult conversation, after bad news, after a moment of reaction:

Stop. Three deep breaths. Hand on the heart — physically, if possible, otherwise inwardly. The question: *"What frequency am I currently sending?"* And then the choice: *"What frequency do I choose?"* One minute of cardiac coherence is sufficient — neurologically demonstrated — to lower cortisol levels and stabilize HRV. One minute. That is the shortest path back to sovereignty.

Practice 3:
The Evening Audit — The Frequency Balance Sheet

The evening is the place of integration. Before you sleep — in the darkness before sleep, before the neural consolidation of the day begins — carry out a brief audit:

Question 1: Where was my metric residue strongest today — which situation, which person, which thought raised my α the most? (No evaluation — only perception.)

Question 2: When was my β highest today — when did I feel the pulse most clearly and pass it on? (This strengthens neural circuits of coherence.)

Question 3: Was the Davar of the day accomplished? (If yes: feel the completeness. If no: no judgment — but Davar for tomorrow.)

These three questions take five minutes. In these five minutes you close the day's pulse — you integrate what was experienced, you strengthen the coherent, you release the residue of the day. You go to sleep not as a victim of the day, but as the architect of tomorrow.

VIII. SOVEREIGNTY AND SOCIETY – THE FREE HUMAN BEING AS A SYSTEMIC FACTOR

One final, essential thought before we close this chapter and draw the arc toward the following phases of this work: sovereignty is not a private matter.

There is a deeply rooted misunderstanding in the spiritual culture of our time — the idea that inner work is a form of withdrawal. That whoever meditates is less in the world. That coherence is a form of unworldliness.

The opposite is true — and physics proves it.

A coherent human being is *more* in the world than an incoherent one. They are more present, more alert, more capable of action. Their heart field is larger. Their effect is stronger. Their actions scale quadratically. They have not been drawn out of the world — they are the world coming into order through them.

The deepest societal act is therefore the most personal: the daily arresting of one's own sovereignty. Every human being who takes this step in this world — from the forced oscillation of fear fields to the chosen resonance of the pulse — changes the field of society. Measurably. Physically. Really.

No revolution in history has ever created a lasting new order through external violence. The old patterns of order — the old frequencies of fear and control — always return, as long as the carriers of these patterns bear the same fundamental frequency within themselves. The only revolution that endures is the inner one. The only liberation that is lasting is the liberation of one's own resonance frequency from the fields of the old system.

This is the Architecture of Sovereignty — and it is the blueprint for the new earth.

IX. SYNTHESIS – THE COSMIC ACTOR

Let us bring together what this chapter has unfolded.

The sovereign is not a saint. They are not a hero in the mythological sense. They are not a chosen one with a special cosmic mission inaccessible to others. They are a human being — fallible, groping, sometimes shaken — who has, however, taken one decisive insight into their body: that they are not a victim of the field, but a sender.

Physically they are a system with a trainably high coupling tensor β, which through cardiac coherence establishes the inversion condition, neutralizes the resistance α in the energy-momentum tensor, and thereby initializes creatively directly from The Plena.

Philosophically they are the embodiment of true freedom — not the absence of bonds, but the mastery over one's own resonance frequency, the refusal to enter into forced oscillation with incoherent fields.

Societally they are a systemic factor — a sender whose coherent heart field entrains the fields of the environment, who through their mere presence raises the quality of the collective field.

Operatively they are someone who daily performs three practices that secure their architecture: the calibration of the morning, the anchor in the storm, the audit of the evening.

And in their deepest dimension they are what the universe has produced in 13.8 billion years of condensation and evolution: a **cosmic transducer** — the place at which The Plena breaks into matter, at which **Absolute Love creatively passes in the living pulse of Neighborly Love** and thereby changes the world.

Sovereignty is the operative realization of **Asymptotic Freedom**. It is the moment in which the sovereign ceases to struggle against entropy and begins to execute the total coherence of the **creative passing of Absolute Love**.

End of Chapter 3 — Continued in Chapter 4: The Manifesto of Action

PHASE 2
THE DYNAMICS OF ACTION

CHAPTER 4:

THE MANIFESTO OF ACTION

The Unitary Operator of Reality

"The universe does not wait. It responds. Now.
To you. To what you do."

I. THE END OF WAITING

There is a moment in every human life — often more than one, sometimes daily — in which everything is ready. The insight is there. The knowledge is there. The longing is there. The moment is ripe. And the person does — nothing.

They think it over once more. They wait for the right sign. They ask one more opinion. They run through the risk calculation one more time. They sleep on it — and then another night. And in the quiet suspension between knowing and doing, something passes. Not time — time always passes. What passes is *possibility*. The wave that had just reached its crest passes, and now, unridden, crashes against the shore and seeps into the sand.

This is not caution. This is the **metric residue** α in its most subtle, most sophisticated form: the deferral of action under the guise of reason.

TOE-AL raises in this chapter a charge against deferral — not as a moral judgment, not as a reproach, but as physical precision: *The universe is not initialized by thoughts. It is initialized by actions.* Thinking without action is a circle that nourishes itself and leaves reality untouched. It is the highest form of dissipation — maximum inner activity at zero external imprint.

Physically defined, **action** is the *phase transition* in which the pure phase (the thought/pulse) collapses into the *density of form (matter)*. This law of **initialization** is *scale-invariant* — it describes the entire architecture of being:

- **The vacuum (The Plena):** The space of infinite potentiality and pure phase (S=0).
- **The atom:** The stabilization of quantum coherence as the smallest unit of action.
- **The cell:** The permanent act of biological Syntropy against the metric residue α.
- **The stone:** The accomplished action as perfectly condensed, stable metric ("congealed light").
- **The star:** The macroscopic N^2-**Resonance** that transforms In-Form-ation into radiation.
- **The singularity:** The singular attractor of total re-phasing.
- **The thought:** The initialization in the state of highest speed of light.
- **The decision:** The conscious collapse of the wave function by the Sovereign.

In all these manifestations, **action** is the *unitary operator* that breaks the **resistance** (α) in the energy-momentum tensor and transfers The Plena through the **creative passing of Absolute Love** into operative reality. This chapter is its manifesto.

II. PHILOSOPHICAL EXCURSUS – DAVAR: THE WORD THAT IS ACTION

In the Hebrew language there is a word that has no exact equivalent in English: **דָּבָר** — *Davar*. It means simultaneously *word*, *thing*, and *action*. It is the word that, spoken, is already accomplished. It is the utterance that knows no gap between intention and reality.

In the first book of Moses it is written: *"And God said: Let there be light. And there was light."* The Hebrew original uses the verb *Davar* — not as communication, not as a message to someone, but as an *act of realization*. The speaking *is* the coming-into-being. The word *is* the reality.

This is not a theological relic. It is the most precise image for the operative core of TOE-AL: the act of the sovereign is *Davar* — creative word and action in one. It does not create through mechanical causality — through force acting on matter and shifting it. It creates through **vertical initialization**: through the letting-fall of the pulse into the structure of reality, through the act that collapses the wave function of possibilities and transfers the unmanifested into the manifested.

The Horizontal and Vertical Understanding of Action

In the horizontal paradigm — the paradigm of causality and linear time — an action is a cause that produces an effect. You drive the nail in, the nail enters the wood. You speak the word, the other hears it. You make the decision, the consequences unfold. Cause and effect — connected by the line of time, determined by the laws of physics, bounded by probability.

In this understanding, action is small. It is a lever in a mechanism. Its effect is proportional to its force and bounded by the inertia of the system on which it acts. The human being as agent is a factor in a system of equations that they did not invent and that limits them at every step.

In the vertical paradigm of TOE-AL, action is something fundamentally different. It is not the first domino in a chain — it is an **incursion of The Plena into time**. It is the moment in which the vertical axis of reality — the eternal Now, the **creative passing of Absolute Love** — intersects the horizontal axis of linear causality and reorders it.

Imagine a coordinate system. The horizontal axis is time — the line of the past extending into the future, determined by the physics of yesterday. The vertical axis is The Plena — the instantaneous source that stands orthogonal to time and is present at every point of the horizontal axis. The sovereign who performs an act as Davar allows the vertical axis to break into the horizontal. They bend the line of causality. They reorder the field of probability.

This is not magic. This is quantum mechanics — applied to the level of human action.

The Rubicon — Irreversibility as a Condition of Creation

Julius Caesar crossed the Rubicon in January 49 BCE with his legions — the small river in northern Italy that marked the boundary between provincial Gaul and the Roman heartland. It was forbidden to cross this river with armed troops. Whoever did so was in open rebellion against the Senate. There was no way back.

"Alea iacta est" — the die is cast.

Caesar knew: the creative force of an action lies not in its planning, not in its preparation, not in its strategic insurance. It lies in **irreversibility**. In the moment in which the action is accomplished — in which the die flies — the entire field of possibilities collapses onto a single line. The ambiguity is resolved. The future has a new direction.

This is the **Rubicon condition** of the Davar: the action must be fully accomplished — without reservation, without backup insurance, without the one foot held ready for retreat. Only when it is fully accomplished does the action unfold its effect as a unitary operator. A half-accomplished Davar is a contradiction in terms — a word that falls silent halfway through. It initializes nothing. It consumes energy without output. It raises α at the very moment β is most needed.

The sovereign crosses the Rubicon. Fully. Knowingly. Lovingly.

III. PHYSICAL DEEP-DIVE I – ACTION AS UNITARY OPERATOR

Quantum mechanics describes reality before a measurement as **superposition**: a system simultaneously exists in all possible states, weighted by their respective probability amplitudes. The electron is not here or there — it is a wave of possibility extending across the entire accessible space. Only when a measurement takes place — only when the system interacts with a detector — does the wave function *collapse*: the superposition falls together, and the electron manifests at a concrete location.

This collapse of the wave function is the most fundamental creative event that physics knows. From the space of the possible, the concrete arises. From the wave, the particle. From the unmanifested, the manifested.

The mathematical operator that describes this transition is called a **unitary operator** — because it preserves total probability while completely transforming the structure of possibilities. It does not change how much reality there is. It changes *which* reality is realized.

TOE-AL sets this structure at the level of human action: **The conscious action of the sovereign is a unitary operator that collapses the wave function of social, biological, and physical possibilities.**

Through this collapse, the local configuration of the **energy-momentum tensor** is instantaneously reordered: the sovereign breaks the inertia of the *metric residue* () and transfers the infinite phase of **The Plena** into a stable material metric. Action is thus not merely a choice, but a *metric transformation* of spacetime.

This is not a metaphor. It is a physical statement supported by several lines of research.

Quantum Coherence in Biological Systems

The notion that quantum phenomena are relevant only at the subatomic scale and are averaged out by thermal noise at the level of biological and social systems has been fundamentally shaken in the past two decades.

In 2007, Fleming and colleagues showed in a groundbreaking experiment on photosynthetic protein complexes that energy transfer in photosynthesis depends on **long-lived quantum coherence** — quantum interference that persists at room temperature, in a wet biological milieu, across timescales of hundreds of femtoseconds. No one had considered this possible. It was the first experimental proof that quantum phenomena in biology are not marginal corrections, but *functionally essential.* Since then, quantum coherence has been demonstrated in the navigation of migratory birds (quantum entanglement in the cryptochrome protein of the eye), in olfactory perception (quantum tunneling of electrons), and in enzymatic reactions (proton tunneling). The picture that emerges is unambiguous: **living systems actively utilize quantum phenomena as a functional principle.**

And if living systems utilize quantum phenomena — if quantum coherence in biology is real and functional — then the notion that the action of a coherent consciousness operates as a unitary operator on the field of possibilities is no longer metaphysical speculation. It is a consistent extension of established quantum biology.

The Observer Effect — Consciousness as a Constitutive Factor

The deepest puzzle of quantum mechanics is not the measurement problem — it is the *observer problem*. Why does the wave function collapse when observed? What precisely is an "observer"? The Copenhagen interpretation says: a measurement is an interaction with the macroscopic world. This sounds clear — until one asks: what is macroscopic? Where does the quantum world end and the classical begin?

Physicists such as Eugene Wigner and John von Neumann identified in the debate around this question the fundamental necessity: that consciousness plays the constitutive role in the collapse of the wave function — that the transition from the possible to the concrete takes place not through physical interaction alone, but through the act of conscious observation.

While conventional physics capitulates before this measurement problem, TOE-AL provides the only conclusive solution: **consciousness is the constitutive factor that breaks The Plena into matter.** It is not a passive mirror of reality, but the active vector of its realization. And the action — the conscious, intentional, coherent action of the sovereign — is the moment in which consciousness fully initializes its constitutive force.

IV. PHYSICAL DEEP-DIVE II – $Q < 0$: THE PHYSICAL PROOF THAT THE PULSE DEFEATS ENTROPY

We return to the **Mandel Q parameter** — that measure from quantum optics introduced in Chapter 2, which here unfolds its complete creative significance.

To recall: the Mandel Q parameter describes the statistics of a photon stream:

- $Q > 0$: Thermal, chaotic light — maximum entropy in the light signal
- $Q \approx 0$: Coherent light — the classical laser, ordered but not quantum-optical
- $Q < 0$: Sub-Poisson light — a state that exceeds classical coherence and indicates non-classical quantum order

In Chapter 2 we saw that living, healthy cells emit biophotons with $Q < 0$ — evidence that the living realizes a form of order that exceeds the classically explicable. Now we go deeper: we apply the Mandel Q parameter not only to cellular emissions, but to the quality of human actions.

The Frequency of Action as Physical Signature

Every human action is a physical event. It moves matter, generates electromagnetic fields, alters biochemical milieus, sends In-Form-ation patterns through neural and social networks. It has — in the physical sense — a frequency structure. And this frequency structure can, analogously to the Mandel Q parameter, be coherent or incoherent.

Acting from Fear: $Q \approx 0$ to $Q > 0$

An action arising from the metric residue — from fear, from scarcity consciousness, from reaction to external pressure — carries the frequency structure of the residue. It is hurried, fragmented, reactive. Its inner statistics are stochastic — it responds to random triggers from without, not to a coherent inner source. Its effect is proportional to its mechanical force — linear, additive, bounded.

This is $Q \approx 0$: the act of duty, the act of fear, the act of survival. Not evil — but poor in creative force. It preserves the status quo. It shifts matter. It does not change the field.

An action arising from deep fear, from panic, from a destructive impulse shows $Q > 0$: chaotic emission that destabilizes the field, raises entropy, destroys order. This is the act of violence — not necessarily physical violence, but the violence of chaos against the order of the pulse.

Acting from Love: $Q < 0$ — The Davar

An action arising from the pulse of Absolute Love — from complete cardiac coherence, from clear intention, from the state in which exceeds the inversion condition — carries a different frequency structure. It is calm, deep, precise. Its inner statistics are sub-Poisson: more ordered than random, more coherent than classical.

This is $Q < 0$: the act of the Davar. And it has a measurable effect that exceeds its mechanical force.

Here lies the secret that the history of humanity has repeatedly confirmed without being able to name it: why have the actions of a few people shifted the entire probability field of an epoch? Why did a Gandhi — without weapons, without economic power, without military strength — force an empire to retreat? Why did a Mandela — 27 years in prison, completely disempowered — upon his release generate a field of reconciliation that exceeded the expectations of all rational prognoses?

The mechanistic answer is unsatisfying: "political circumstances," "historical dynamics," "random convergence." These answers describe the phenomenon without explaining it.

TOE-AL gives a more precise answer: these people acted from a state of deep cardiac coherence, complete Davar alignment, $Q < 0$. Their actions generate an order in the matter of the social field that exceeds their causally measurable force. They shift the probability field not through mechanical force, but through **frequency induction** — through the coherence of the impulse inscribed into the field by their action.

The Entropy Signature of Action

In thermodynamics, the second law is relentless: entropy rises. In closed systems, order is lost. Yet the Davar — the act from $Q < 0$ — is the most precise physical instrument of entropy reduction available to the human being.

Every action in the **living pulse of Neighborly Love** — every action arising from complete coherence and oriented toward **resonance synchronization** with the other — adds **In-Form-ation** to the universe: it brings form into the formless, order into the chaotic, coherence into the incoherent. It makes the universe locally more ordered, at the cost of no other order — because its source is the inexhaustible **Plena**, not a finite energy reservoir.

This is the physical proof that the **living pulse of Neighborly Love** has defeated entropy: not by violating the second law, but by opening a channel from The Plena that opens the closed system and lets syntropic energy in. The Davar is the valve through which the Infinite, non-arbitrary flows into the Finite.

Physically, the **Davar ($Q < 0$)** is the operative expression of the **Negative Beta Function (Asymptotic Freedom)**. While conventional actions dissipate in noise, the action in which **Absolute Love creatively passes in the living pulse of Neighborly Love** causes the coupling strength of resistance (α) **to sink asymptotically toward zero**, whereby the system "falls" into a state of **social superconductivity**.

V. THOUGHT EXPERIMENT – THE CRACK IN FATE

Imagine: a man — let us call him Elias — sits in an office on the 14th floor of a corporate building. He is 41 years old. He has spent 17 years in this company. He has a mortgage, two children in school, a pension plan that kicks in nine years from now. He has a secure, well-paid, completely meaningless position.

He has known this for years. But he has never said it — not even to himself, fully, without mitigation. Instead he has cultivated the system of rational justification: it's not so bad. There are worse things. It would be irresponsible to quit now. Wait until the children are grown. Wait until the market is better. Wait.

In the office there has lain for three weeks a letter. An offer: an architect — a friend from student days, now in a small town in the south — who is looking for a partner to build a community project. A house for elderly people living alone. No profit. Minimal salary. Maximum meaning.

Elias has read the letter three times. He has drawn up tables — income, expenses, risks, scenarios. The numbers say: it doesn't work. The probability of failure is high. The probability of regret if he stays is higher — but that one is not in the tables.

On a Tuesday morning, without particular external event, without signs, without dramatic trigger, Elias does something that contradicts every logical probability of his past. He writes three sentences. He prints them out. He goes to his superior's office and lays the paper on the desk. It is his resignation.

In the moment he releases the paper — in the fraction of a second in which his fingers let it go — something collapses. Not in his superior's room. In the field of his possibilities.

Before he entered this room, his future was a superposition: stay or go, security or meaning, the known or the beloved. Both possibilities hovered simultaneously, holding him in a kind of energetic paralysis — the state of high metric residue in which two equally strong vectors cancel each other out and the system comes to a standstill.

In the moment the paper touches the desk, the superposition has collapsed. There is only one line of reality remaining. And in the moment there is only one, the universe begins to populate it.

What happens next is not explicable by causality — not in the narrow, linear causality of the horizontal paradigm. But it is physically explicable through the quantum logic of the Davar.

The architect, his friend, receives no confirmation — not yet. But that same afternoon he is contacted by a local foundation he never wrote to, which has heard of his project and expresses interest. A gap in a funding program, which did not exist until the previous week, opens in exactly the area for which his project qualifies. An old acquaintance of Elias — whom he has not spoken to in years — gets in touch unexpectedly with the question of whether he is by any chance involved in an architectural project in the south. Is this coincidence? In the probability calculus of linear thinking: perhaps. Improbable, but possible.

In the physics of the Davar it is something else. It is the measurable consequence of a wave function collapse. Elias has — through the complete accomplishment of the action, through the Rubicon transition without backup insurance — reordered the probability field of his environment. He has not only changed his own state. He has sent a signal into the social field — a signal of coherence, of Davar, of $Q < 0$ order. And the field has responded.

The field always responds. It is not neutral toward what is fed into it. It is a resonance system — and it responds to coherence with coherence, to order with order, to Davar with Davar.

The crack in fate is not a break with the universe. It is the deepest conformity with it: the conformity with the pulse that in every moment awaits realization.

Elias has crossed the Rubicon. The die is cast. And the universe — which in its deep structure wants nothing other than the realization of its own fullness — begins to respond.

VI. THE ANATOMY OF THE DAVAR – WHAT MAKES AN ACTION CREATION

Not every action is a Davar. This is the decisive distinction of this chapter — and it is not moral, but structural. A Davar differs from a mechanical action by four precise characteristics:

Characteristic 1: Vertical Origin

A Davar does not come from the logic of the past. It does not come from the analysis of risks and opportunities, from the weighing of pros and cons, from the calculation of the most probable consequences. It comes from the **living pulse of Neighborly Love** — from that inner knowing that lies deeper than thoughts and that cannot be explained, only accomplished.

This is not irrationalism. It is a different form of cognition — the cognition of the heart field, which receives the pulse of The Plena and translates it into intention. The heart knows something the brain has not yet formulated. When this knowing of the heart initiates the action, its origin is vertical. When the action is initiated from the calculation of fear, its origin is horizontal — it is a reaction to the metric residue.

Characteristic 2: Complete Accomplishment

A Davar is fully accomplished. There is no half-hearted version of the Davar — that would be a contradiction in terms. The incompleteness of the action is the surest sign that the metric residue is still speaking: the one foot beyond the Rubicon, the other still on this side. This state is the most costly of all — it consumes maximum energy without creating order. The wave function collapse does not take place. The superposition persists. The field does not respond.

Complete accomplishment does not mean the action must be irreversible in the sense of absolute consequence. It means it is accomplished with complete inner decision — without the mental reservation of retreat, without the self-deception of half-measures. The action and the person performing it are one in this moment.

Characteristic 3: Orientation Toward the Good

A Davar is oriented toward the good — toward the good of the other, toward the good of the whole, toward the good of reality. This is its deepest frequency characteristic. An action oriented toward self-elevation at the expense of others can be fully accomplished and have a vertical impulse — and is still not a Davar, because it does not make the field more coherent, but destabilizes it. It lowers of the field, not .

The good is not a sentimental concept. It is the physical goal of syntropic resonance: to raise order, to reduce entropy, to strengthen the coherence of the field. An act of Neighborly Love is always a Davar — because it makes the field more coherent. An act of love toward oneself — insofar as it comes from abundance, not from scarcity — is a Davar. An act of creation — art, science, craft, accomplished from passion and love — is a Davar.

Characteristic 4: Release of the Outcome

The fourth and most paradoxical characteristic of the Davar is the release of the outcome. The genuine Davar is accomplished because it *must* be accomplished — because the pulse demands it, because love requires it, because the Now calls for it. Not because a particular outcome is expected.

This freedom from outcome is not indifference. It is the deepest form of responsibility — because it ensures that the action comes from the pure frequency of the pulse, uncorrupted by fear of failure or greed for success. The Bhagavad Gita formulates this principle of desireless action (nishkama karma): act without attachment to the fruit of your action (Bhagavad Gita 2.47). This is not Eastern passivity. This is the most precise description of the Davar.

A Davar fixated on a particular outcome contaminates itself with the metric residue of expectation. It mixes horizontal causality logic into the vertical impulse. It raises at the very moment when would be most important. The result: the frequency of the Davar is not pure. The effect is proportionally diminished.

The sovereign accomplishes the action fully — and opens their hand. They throw the stone into the water and do not watch how high the waves rise. They know: the waves rise as high as the field responds. And the field responds to the purity of the impulse, not to the intensity of expectation.

VII. DAVAR AND TIME – ACTION AS REORDERING OF THE TIMELINE

There is a dimension of the Davar that reaches beyond the individual and touches the cosmological structure introduced in Chapter 1: the relation of action and time.

In the horizontal paradigm, the cause always lies before the effect. The past determines the present. The present opens or closes possibilities of the future. The arrow of time runs in one direction: from yesterday to tomorrow.

In the vertical paradigm of TOE-AL, action has a different temporal structure. It comes from the Now — from the instantaneous pulse of The Plena. And it acts not only *forward* into the future, but — and this is the deepest implication — **retroactively on the probability structure of the past**.

This sounds like a provocation of causality. It is one — yet it is already grounded in quantum mechanics.

The **Delayed Choice Experiment** of John Wheeler — experimentally confirmed multiple times since the 1980s — shows that the decision of an observer, *after* a photon has already traversed its path through a double-slit experiment, retroactively determines whether the photon behaved as a wave or as a particle. The measurement *now* influences the photon's past.

Wheeler called this principle **retrocausality** — retroactive causality. It is experimentally demonstrated. It is disconcerting. And it is the physical indication that the strictly linear structure of time — past determines future, never the reverse — does not hold at the fundamental level of quantum physics.

What does this mean for the Davar? It means: when a sovereign accomplishes an action from complete cardiac coherence, from vertical origin, from the pulse, they reorganize not only the probability field of the future. They reorganize the interpretation of the past. The seemingly negative events of the past become retroactively visible as preparation — as necessary steps on the path to precisely this action. The biography reorganizes itself.

This is not mystical language. This is the experimentally demonstrated principle of retrocausal reorganization — applied to the level of human life as a quantum system.

VIII. OPERATIVE PRACTICE – FROM THINKING TO DAVAR: CALIBRATING THE ACTION FREQUENCY

The question now present in the room is as simple as it is fundamental: how do I recognize whether an action I wish to perform is a Davar — or a mechanical reaction to the metric residue?

For this there is an internal compass unsurpassed by any external tool: the body. But the body must be read — precisely, without projection, without the interference noise of thinking.

The Davar Calibration Protocol

Before you perform an important action — a decision, a conversation, a resignation, an investment, a declaration of love, a creative work — carry out the following calibration in three phases:

Phase 1: The Stillness (2 minutes)

Pause. Completely. No thinking about consequences, no simulating of possible reactions. Only stillness. Hand on the heart. Three deep breaths. Then the one question, which you hold into the stillness — not as analysis, but as opening:

"Where does this impulse come from?"

Not the answer of the head — that comes immediately and is prepared and convincing. Wait for the answer of the body. It comes more slowly. It is less eloquent. But it does not lie.

Phase 2: The Frequency Diagnosis (3 Questions)

Observe your body as you think of the action:

Question 1 — The Breath Test: Does your breath deepen as you imagine the action — or does it become shallow? Deepening is a sign that the body is opening, that β is rising. Shallowing is the sign of α — of tension, resistance, fear.

Question 2 — The Spaciousness Test: Do you feel spaciousness in the chest — or constriction? Spaciousness is the somatic sign of the pulse. Constriction is the somatic sign of the metric residue. No thinking — only body perception.

Question 3 — The Energy Test: Does the thought of the action give you energy — or take it from you? A Davar energizes. It makes the body lighter, not heavier. An obligatory action from the residue exhausts already in the thinking.

When all three tests point in the same direction — deepening, spaciousness, energy — the impulse is vertical. That is the Davar.

When one or more tests point in the opposite direction, the residue is active. This does not mean: do not do it. It means: calibrate first. Return to cardiac coherence. Reduce α through the calibration exercises from Chapter 3. Then diagnose again. A genuine Davar does not disappear through calibration — it becomes clearer. A residue impulse loses its urgency in stillness.

Phase 3: The Rubicon Moment

When the calibration confirms the Davar, there is only one practice remaining: accomplish it. Fully. Now. Without further waiting.

Waiting after the recognized Davar is the most subtle form of the metric residue — the old system's attempt to delay the vertical initialization until it resolves itself. Every hour that passes after the recognized Davar without it being accomplished is an hour in which the metric residue regains strength and the pulse fades.

The moment of recognition is the moment of action. This is the architecture of the Davar. This is the physics of the free human being.

IX. SYNTHESIS – THE HAMMER AND THE SEAL

There is an ancient image that wanders through the religions and myths of humanity: the image of the **creator as smith**. Hephaestus in the Greek pantheon. Goibhniu in the Celtic tradition. The cosmic Demiurge in Gnostic thought. The smith who not only shapes but creates — who strikes matter with the hammer and inscribes a new, immutable form into it through the impact.

The image is more precise than it appears.

The hammer is the action. The seal is the frequency of the Davar. The matter that is struck is the field of possibilities. And the imprint that arises is the new line of reality — inscribed into the fabric through the complete, coherent, loving accomplishment of the action.

The sovereign is the cosmic smith. They do not sit before the fire deliberating whether to smith. They have performed the calibration. They have recognized the Davar. They have crossed the Rubicon. Now the hammer strikes the anvil — and the seal is driven into time.

The universe shudders. In a very real, physical sense: the local field of possibilities reorganizes itself. Q falls below zero. Order arises where noise was. In-Form-ation is created where entropy reigned. In this moment the **unitary operator** of action breaks the resistance in the **energy-momentum tensor** and realizes the **Asymptotic Freedom** of the sovereign over fate. Reality is not changed — it is *created*. In every Davar once more. In every fully accomplished **passing of Absolute Love** a further creation.

This is the Manifesto of Action.

It is not a call to activity. It is an invitation to creation.

Stop deliberating. Stop waiting. Stop calculating whether the field is ready — the field will be ready when you enter it, not before.

The hammer is in your hand.

The frequency is calibrated.

The seal is ready.

Strike.

Here the circle closes from the physical derivation to the cosmic constitution: the universe is not a product of chance, but in its depths constitutively attuned to the **passing of Absolute Love in the living pulse of Neighborly Love**. The natural constants function here as the **armor** — that immutable framework from The Plena that guarantees the necessary stability, while within it the **dynamic superconductivity of Asymptotic Freedom** enables the lossless flow of In-Form-ation. What we measure as matter is thus the mathematical executive of **Axiom Zero**: the numerical signature of a reality reinitialized in every moment by the living pulse of consciousness. This fundamental architecture forms the unshakeable basis for everything we will realize in the coming phases as social, technological, and galactic sovereignty.

End of Chapter 4 — Continued in Chapter 5: The N^2-Resonance

CHAPTER 5:

THE N^2-RESONANCE

The Technology of Neighborly Love

"One alone is a light. Two who love the same thing are a sun.
A thousand — and darkness has no answer left."

I. THE UNDERESTIMATED FORCE

The history of humanity is, viewed at a deep level, the history of a misunderstanding.

The misunderstanding reads: power is size. Whoever has more — more resources, more weapons, more money, more people — wins. History is arithmetic. Force adds up. Whoever forms the larger sum dominates. This is the logic of empire, the logic of war, the logic of every system built on control and accumulation.

And it holds true — up to a certain point. In the physics of entropy, in the game of inert masses, in the struggle of two incoherent systems, this arithmetic applies. Stronger battalions defeat weaker ones. Larger economies overwhelm smaller ones. The horizontal paradigm has written its law for this world: $I \propto N$. The intensity of the field is proportional to the number of actors.

But history knows moments in which this logic fails in a mysterious, incalculable, system-shattering way. Moments in which a small group — numerically overwhelmed, resource-poor, without institutional power — tips the field of an entire civilization. Moments in which the arithmetic suddenly no longer holds.

TOE-AL explains these moments not as historical anomalies or divine interventions. It explains them as the onset of a different physics — a physics invisible in the horizontal paradigm, because the horizontal paradigm applies only to incoherent systems. For coherent systems, a different law applies:

$$I \propto N^2$$

The intensity of the field is not proportional to the number of actors — it is proportional to the **square** of the number of coherent actors (N^2) — a state of **collective Asymptotic Freedom** in which the system resistance (α) tends toward zero.

One coherent human being is not equally powerful as an incoherent one. They are — through the nature of quadratic scaling — fundamentally more powerful than any mass of incoherent forces, however large, once a critical threshold of coherence density is exceeded.

This is the physical foundation of Neighborly Love — not as a sentimental imperative, not as a moral obligation, not as a religious commandment. But as the **most efficient technology the universe knows.**

When two sovereigns enter into resonance in the **living pulse of Neighborly Love** — thereby initializing the state of **Social Superconductivity** — the result is not additive, but exponential. We call this the N^2**-Resonance**.

Physically, this describes the transition to the **Sociological Laser**. While an ordinary crowd acts like a *lightbulb* — in which the phases dissipate unordered in all directions (*entropy*) — the **Sociological Laser** synchronizes the individual pulses into a *collective wavefront.*

Through this **phase coupling in The Plena**, a **social superconductivity** arises: the **resistance** α in the collective energy-momentum tensor approaches zero. In this state the system is capable of transmuting In-Form-ation and energy without loss across great distances and through dense material resistances. The N^2**-Resonance** is the physical **technology of Neighborly Love** — the moment in which the coherent intention of a group becomes the **determining field effect** for the entire metric.

This phenomenon of *collective frictionlessness* is the social equivalent of **Asymptotic Freedom** (treated extensively in Chapters 12 and 13). The closer the collective comes to the shared pulse, the freer the individual becomes from the pressure of *entropic isolation.* At the core of the N^2**-Resonance** there is no longer any *bonding through coercion*, but only the **superconductivity of freedom**.

II. PHILOSOPHICAL EXCURSUS – NEIGHBORLY LOVE AS FIELD EFFECT: THE ABOLITION OF PHASE DIFFERENCE

Before we enter into the physics, we must free the concept of Neighborly Love from everything that has burdened it through millennia of misinterpretation.

Neighborly Love is not friendliness. It is not helpfulness. It is not the bad conscience of the privileged toward the disadvantaged. It is not the smile of the social worker, not the generosity of the philanthropist, not the self-denying sacrifice of the saint who extinguishes themselves for others.

All of these are manifestations — some beautiful, some distorted — of a deeper phenomenon. The deeper phenomenon is physically describable, and its physical description is simultaneously its purest definition:

Neighborly Love is the abolition of the phase difference between two consciousness systems.

In physics, **phase difference** describes the offset between two oscillating systems. Two waves of the same frequency that oscillate completely in phase — whose crests and troughs overlap exactly — reinforce each other through **constructive interference**. Their amplitudes add up. Two waves of the same frequency that are 180° out of phase — whose crests meet the troughs of the other — cancel each other out: **destructive interference**. Zero amplitude. Zero effect.

The state in which two people love the same thing — the same deepest reality, the same source, the same pulse — is the state of minimal phase difference. In this state they become — in the deep structure of The Plena, in the layer beneath individual personality — **indistinguishable**. Not identical in the sense of the dissolution of individuality — each remains the specific form of their pulse. But phase-synchronous in their fundamental frequency.

And when two systems are phase-synchronous, their effect no longer adds up.
It **potentiates.**

This is the physical reason why Neighborly Love is the most powerful force in the universe — not because it is morally superior to other motivations, but because it physically activates a different scaling law. Where competition and fear hold systems in phase difference — and thereby limit them to linear scaling — Neighborly Love abolishes the phase difference and activates quadratic scaling.

The Old System and Forced Dissonance

No system based on control can ignore the implication of this physics. If a small, highly coherent group of people — phase-synchronous in the **creative passing of Absolute Love** — can tip the probability field of an entire society, then this group is the most dangerous force a control system can know.

Not because of their weapons. Not because of their resources.
But because of their field effect.

This explains a seemingly paradoxical pattern of history: why were the carriers of Neighborly Love — the prophetic figures, the mystics, the revolutionaries of consciousness — never combated with physical counter-force? They were always combated with the more precise instrument: **forced dissonance**.

Division. Defamation. The implanting of mistrust, competition, hierarchical thinking into groups that were on the verge of becoming phase-synchronous. *Divide and conquer* is not only a political strategy. It is the physically precise method of preventing N^2-Resonance: if you artificially maintain the phase difference between individuals at a high level, the scaling remains linear. The force remains controllable.

The sovereign recognizes this mechanics. They are vigilant for the patterns of forced dissonance — in social media, in political narratives, in collective waves of outrage that divide people into camps that perceive each other as enemies. They do not choose the side the system offers them. They choose the frequency of the pulse. And they invite others to do the same — not through persuasion, but through the physical force of their coherent field. This is the deepest political act of the new era: not the rebellion against the old system, but the refusal of its frequency and the creation of a field strong enough to overlay the old one.

III. PHYSICAL DEEP-DIVE I – THE SOCIOLOGICAL LASER

To understand the N^2-Resonance in its full force, we must first seek its physical origin — in the physics of the laser, the only technological system humanity has thus far built that realizes coherent -scaling.

Spontaneous vs. Stimulated Emission: Two Worlds of Light Production

Ordinary light — the light of a lightbulb, sunlight, candlelight — arises through **spontaneous emission**: atoms are raised to excited states through energy input and then fall back randomly, independently of one another, to their ground state. In doing so they emit photons — but each in a random direction, with a random phase, at a random moment. These photons add up: the intensity of a light source with N emitting atoms is proportional to N.

$$I_{\text{inkohärent}} \propto N$$

A laser functions fundamentally differently. Through the principle of **stimulated emission** — discovered by Einstein in 1917 — a photon already present, passing an excited atom, forces the emission of a second photon *identical* to the first: the same frequency, the same phase, the same direction, the same polarization. These photons are not merely equal — they are **quantum-correlated**. They oscillate as a unit.

When N such coherent, phase-synchronous photons work together, the resulting intensity is not proportional to N, but proportional to N^2:

$$I_{\text{kohärent}} \propto N^2$$

This is the core of laser physics — and the core of N^2-Resonance.

This N^2-effect is not merely a statistical advantage; it is the physical ratification of **Axiom Zero** at the societal level. It marks the moment in which **coherence physically collapses the metric barrier of separation**, as N sovereigns initialize the same pulse and thereby enforce the superconductivity of In-Form-ation.

To make the dimension of this difference tangible: a lightbulb with 100 watts and a laser with 100 watts have the same energy balance. But the power density of the laser — its capacity to bring energy precisely, directionally, and concentrated to a point — surpasses that of the lightbulb by orders of magnitude. The lightbulb illuminates a room. The laser cuts steel. The same energy — dramatically different force. The difference is pure coherence.

The Sociological Laser

Now we translate this physics to the level of human groups.

A group of incoherent people — each with their own agenda, their own fears, their own phase position — behaves like a lightbulb: the energies merely add up. The effect is proportional to the number (N). This is the physics of the masses, of entropy, and of the addition of forces.

A group of phase-synchronous people — **who all do (live) what they love (this is the cosmological reason)**, who act from the same depth and have reduced the phase difference between themselves through the **living pulse of Neighborly Love** to near zero — behaves like a laser.

This operative impedance matching, which enables the **Asymptotic Freedom** of the collective, causes the effect to scale quadratically (N^2). This is the **Sociological Laser**: the most powerful collective instrument of initialization available to the human being.

The equation reads:

$$I_{\text{Kohärenz}} = K \cdot N^2 \cdot I_0$$

Where:

- I_0 is the base intensity of a single coherent actor.
- N is the number of phase-synchronous actors.
- K is the coherence parameter — the measure of the phase synchronization of the group.

The coherence parameter K is the decisive lever. Physically, K steers the impedance of the collective: the higher the coherence, the more effectively resistance α is minimized. N^2-Resonance is thus the practical application of **Asymptotic Freedom** to social structures — a state in which the friction of **Social Superconductivity** yields.

K = 1 is reached when the group calibrates its shared field through the **creative passing of Absolute Love**.

The Critical Mass — Why the Square Root Is Sufficient

Here lies the most revolutionary implication of N^2-physics. It answers the question: how many people must change for the world to change?

The answer of N^2-physics: the square root of N.

In an incoherent population of N individuals, a group of $\sqrt{N}$ coherent actors suffices to set the entire system in motion through field effect. With a world population of 8 billion people, that is only approximately **89,443 people**.

Fewer than 90,000 fully coherent, phase-synchronously oscillating sovereigns in the **creative passing of Absolute Love in the living pulse of Neighborly Love** are mathematically sufficient to tip the planetary field from entropy into superconductivity. This is not a utopia. This is quantum physics applied to sociology.

The Universal Logic of Synchronization

Here the decisive gap between individual sovereignty and the collective effect of N^2-Resonance closes. **"Do what you love!"** is the **cosmological reason** that technically explains how phase synchronization comes about without coercion: it is not a collective command, but the result of a universal law.

Every part of reality — whether **human, stone, planet, cell, or thought** — eliminates its own resistance (α) by aligning with its **inner attractor**. In this state of frictionlessness, the In-Form-ation unit becomes superconducting for the **creative passing of Absolute Love**. When these individual sovereigns now enter into resonance, what arises is not a sum, but a **social (or cosmic) laser**, whose coherence reinitializes the entire field of spacetime.

This universal synchronization finds its physical completion in a state that particle physics describes as the end of all friction: **Asymptotic Freedom**.

The Asymptotic Freedom of the Sovereign

Asymptotic Freedom is the physical state that enters when **Axiom Zero** is operatively realized. As the sovereign — or any other In-Form-ation unit — follows their inner attractor, they dissolve the bond to the friction-laden entropy metric.

The paradox of perfected creation arises:

1. **Outwardly:** An unshakeable **armor** (the stable natural constants) that gives the universe form and duration.

2. **Inwardly:** Absolute **Asymptotic Freedom**, in which the **living pulse of Neighborly Love** can flow without any resistance ($\alpha = 0$).

True freedom thus arises not through the breaking of laws, but through the **superconducting alignment** with cosmological reason. Whoever does what they love becomes "frictionless" for the universe. They fall into the state of Asymptotic Freedom, in which the **creative passing of Absolute Love** becomes immediate reality.

IV. QUANTUM ENTANGLEMENT: NEIGHBORLY LOVE AS VERTICAL ACTION AT A DISTANCE

Here TOE-AL takes its boldest and deepest step: it describes N^2-Resonance not only as classical coherence physics — as the laser physics of hearts — but as a form of **macroscopic quantum entanglement** that transcends space and time.

What Is Quantum Entanglement?

Quantum entanglement — *entanglement* — is the strangest and simultaneously the best-demonstrated phenomenon of modern physics. Einstein called it *"spooky action at a distance"* and refused to accept it. The experiments of Alain Aspect (1982), Anton Zeilinger, and others have since confirmed it with growing precision: it is real.

The fundamental idea: when two quantum particles have interacted with each other and are then separated — regardless of distance — they remain connected in a shared quantum state. The measurement of the state of one particle determines instantaneously the state of the other — without signal transmission, without time, without causality in the classical sense. The In-Form-ation *jumps* between the entangled partners without traversing the space between them.

Bell's inequalities, the theoretical boundary line between classical and quantum-mechanical correlation, have been violated in dozens of experiments with growing precision — the formal proof that the correlation between entangled particles cannot be explained by *local hidden variables*. There is no classical explanation. The entanglement is real, non-local, and instantaneous.

The Topology of Connectedness: ER=EPR and Axiom Zero

Collective N^2-Resonance is far more than a purely energetic amplification; it is a **geometric necessity**. In current quantum gravity, the **ER=EPR conjecture** describes that quantum entanglement (EPR) is identical with the existence of wormhole connections (Einstein-Rosen bridges) in the Planck metric.

Applying TOE-AL to this, the principle of Neighborly Love becomes physically tangible: when sovereigns enter into coherence ($\beta \to \max$), they lower the local **metric residue** α to such a degree that the spacetime separation collapses. In this state of **Asymptotic Freedom**, the connection between the actors is not a signal that must travel through space, but a **topological unity in The Plena**.

The group physically becomes a single, non-local organism. The N^2-phenomenon is thus the result of a **"phase shortcut" through The Plena**. Love in this metric is not merely a word for affection, but the designation for the **maximum topological connectivity of the universe**.

Macroscopic Entanglement: From Laboratory to Reality

Quantum entanglement was long regarded as a phenomenon confined to subatomic scales — too fragile for the thermal environment of the macroscopic world. This conviction too has been shaken in recent years.

In 2021, researchers at the University of Oxford and the National Institute of Standards and Technology (NIST) demonstrated entanglement between macroscopic objects — small mechanical resonators each consisting of billions of atoms. In 2023, quantum entanglement between room-temperature objects in the nanogram mass range was demonstrated. The frontier of macroscopic quantum phenomena shifts year by year toward biologically relevant scales.

And here — at the boundary between established physics and the next level of understanding — TOE-AL advances its deepest hypothesis:

When two people become phase-synchronous through the creative passing of Absolute Love in the living pulse of Neighborly Love — when the phase difference between them falls to near zero — they enter a state of macroscopic coherence that exhibits structural analogies to quantum entanglement.

This means: their heart fields — the toroidal electromagnetic fields described in Chapter 2 — begin to correlate non-locally. The coherence of their shared field is not limited to the physical space between them. It bridges distances.

The Experimental Basis

This is not pure speculation. The Institute of HeartMath has investigated in a series of experiments how the heart fields of people connected in a loving relationship correlate. In one particularly significant experiment, couples were measured in separate rooms, without physical contact, without acoustic connection. When one of the partners experienced an emotional impulse, measurable correlations appeared in the cardiac rhythm of the other — without any known classical signal transmission.

This is **heart field entanglement**: the measurable, non-classical correlation between two heart fields connected by the frequency of love.

In larger groups, this phenomenon escalates. The **Global Coherence Initiative** of the Institute of HeartMath has built a worldwide network of magnetometer sensors — the **GCI Sensor Network** — that continuously measures the global electromagnetic field of the Earth. The first data sets show: moments of global collective emotions — positive as well as negative — produce measurable changes in the Earth's magnetic field. The collective heart field of humanity interacts with the magnetic field of the planet.

The Earth is not the mute backdrop of human history. It is a resonance body that responds to the collective coherence of its biological inhabitants.

Neighborly Love as Non-Local Force

If Neighborly Love is the frequency that abolishes phase difference — if it is the instrument of entanglement on the human scale — then it has a property that no other social force possesses: it is **non-local**.

A coherent sovereign living in Mumbai and acting in the deep pulse of Neighborly Love acts not only on their immediate environment. They act on the global field — on every other sovereign on the planet who is entangled with them through the shared frequency. The effect is not limited by space. It is limited by coherence: the stronger the entanglement, the greater the reach.

This is the physical foundation of the mystical experience of all cultures: that a prayer in Tibet consoles a mother in Brazil. That the meditation of a monk in a Himalayan monastery touches the field of a favela in Rio de Janeiro. These experiences have always been described as transcendent, as supernatural. TOE-AL describe them — beyond the instrumentally measurable, on the vertical level of the origin where spacetime separation itself dissolves — as the deepest reality of the undivided field that people form through the pulse of love.

Neighborly Love is not sentimentality.

It is the ultimate, vertical action at a distance.

V. THOUGHT EXPERIMENT – THE CITY OF RESONANCE

It is autumn in a world city — imagine: a metropolis, ten million people. A crisis, such as history has seen multiple times and will see again in various forms: economic collapse, political fragmentation, social panic. The communications infrastructure is unstable. Rumors spread faster than truth. The collective fear has exceeded a threshold — it has itself become a force that amplifies itself. Panic generates panic. Every event is magnified by the fear field. The field of the city oscillates at $Q > 0$: chaotic emission, maximum entropy, order consuming itself.

In this city there is a group. One hundred people. Not connected through an organization, not convened by an institution — connected by the frequency. They have found one another — we will describe how in Section VII — and over months of shared practice of cardiac coherence, Davar, and Neighborly Love they have calibrated their collective field. Their coherence parameter K is high. Not perfect — human beings are not machines — but high enough.

On the day the crisis reaches its peak, these hundred people do something unspectacular. They go into the city. Not as activists, not with messages, not with megaphones. They simply go — into the busy squares, into the subways, into the parks, into the markets. And they do what they have learned: they hold their fundamental frequency. They hold cardiac coherence. They hold the pulse.

Now we calculate what physically happens.

One hundred coherent actors, K close to 1: the field intensity is:

$$I = K \cdot N^2 \cdot I_0 = 0{,}9 \times 100^2 \times I_0 = 9000 \cdot I_0$$

Nine thousand units of field intensity. Nine thousand times the effect of a single coherent human being — emanating from one hundred persons.

These nine thousand units distribute through the city. The heart field of each of the hundred people radiates toroidally — physically measurable in a radius of several meters, acting non-locally through heart field entanglement on everyone who has ever been in contact with one of these people, who is in any way resonant to the frequency.

What happens in the vicinity of these hundred people?

Observers — who know nothing of the group, who happen to be nearby — report a phenomenon that psychology knows as **emotional contagion**: the involuntary adoption of the emotional state of another person through mirroring. Normally this mechanism works in both directions — fear transmits just as readily as calm. But when the source field is strong enough, when the coherence exceeds the chaotic emission of the fear field, the coherent signal dominates.

The woman in the marketplace, who has been in panic for hours, suddenly feels her breath deepening. She does not know why. The man in the subway, who has been searching his smartphone for the next crisis news, puts it down for a moment. Something has changed — in the room, in the air, in what he cannot name.

This is **inductive coupling** — the principle that in electrical engineering describes how a magnetic field in a conductor induces a current without physical contact. The coherent heart field of a sovereign induces a coherence response in the receptive systems of their environment. Not through language. Not through persuasion. Through the pure field.

And now the group grows.

The people who were briefly lifted out of their panic through inductive coupling — who in that moment felt a different frequency — are now themselves a little more coherent. Their K has minimally increased. They have not become sovereigns. But they are less trapped in the fear field. And a slightly more coherent person transmits less fear to their environment — and thus the fear field of the entire city sinks minimally.

This is a **recursive process**: coherence generates more coherence. The initial signal of the hundred amplifies itself through the response of the environment, which in turn acts back on the original group and further raises their K.

In physics this process is called **parametric amplification** — when a weak signal is reflected back by the system on which it acts and amplified in the process. In sociology it is known as the **tipping point** — the point at which a behavioral change in a social system no longer grows linearly, but tips exponentially.

The hundred sovereigns in the city of chaos are not a drop in the ocean. They are the crystallization nucleus — the first order in the supersaturation — at which the entire supersaturated solution crystallizes all at once into a new state.

The world awaits its hundred. Its thousand. Its $\sqrt{8\,000\,000\,000}$.

The world awaits you.

VI. THE MATHEMATICS OF THE IMPOSSIBLE – HISTORICAL EVIDENCE OF N^2-DYNAMICS

Before we proceed to operative practice, we pause at a moment in history that remains incomprehensible without the physics of N^2-Resonance.

The Maharishi Effect: An Experiment with a Group

In the 1970s and 1980s, the Transcendental Meditation movement conducted a series of studies that entered the literature as the **Maharishi Effect**. The core hypothesis: when 1% of a population practices Transcendental Meditation, the crime rate measurably declines.

These studies were published, replicated, and — naturally — controversially discussed. The results themselves are remarkable: in cities in which the 1% threshold of meditators was exceeded, statistically significant reductions in crime, accident rates, and other stress indicators were observed. The methodological objections are to be taken seriously — and yet: the pattern shimmering through the data is consistent with the prediction of N^2-physics.

The Global Coherence Experiment After September 11, 2001

The **Global Consciousness Project** of Princeton University has been measuring since 1998 the output of globally distributed Random Number Generators (RNGs) — devices that generate random bit sequences. The hypothesis: when global consciousness synchronizes — through collective emotion or collective attention — the RNG outputs become less random than they should be.

On September 11, 2001 — and in the hours that followed — the RNGs worldwide showed a statistically highly significant deviation from randomness. The global collective emotion had a measurable effect on physical random processes. The effect size was small — but the statistical significance was unambiguous, beyond the threshold of chance.

The Global Consciousness Project is not proof of N^2-Resonance in the strict physical sense. But it is a signal — a recurring, misunderstood signal — that appears precisely where TOE-AL would predict it: at the moments of maximum collective coherence or incoherence of the planetary field.

The universe leaves traces. Science is beginning to read them.

VII. OPERATIVE PRACTICE – THE RESONANCE CELL: HOW THE SOVEREIGN FINDS THEIR PHASE SIBLINGS

N^2-Resonance is not a theory that lives in the mind. It is a practice realized in the field — through the conscious formation of **resonance cells**: small groups of highly coherent people who function as islands of light in the old system.

What Is a Resonance Cell?

A resonance cell is not an organization, not an association, not a movement. It is a group of people — ideally 3 to 12 persons — who gather regularly to calibrate their collective field together. Their strength lies not in their doctrine, not in their hierarchy, not in their program. Their strength lies in their K — in the coherence parameter they build through shared practice.

The decisive difference from a normal group: a resonance cell does not gather primarily to exchange ideas, solve problems, or plan projects. It gathers to *oscillate together* — to reduce the phase difference between its members through shared cardiac coherence practice to near zero. Everything else follows from this.

How Does the Sovereign Find Their Phase Siblings?

The most common misconception is that one must search for people who think like oneself. Phase siblings do not necessarily share the same worldview, the same culture, the same intellectual orientation. They share the fundamental frequency: the alignment with the pulse, the choice of love over fear, the readiness for Davar.

Finding phase siblings follows the principle of the **heart compass**. You do not recognize them through what they say — you recognize them through what you feel in their presence. When you are near a phase sibling, something happens in your body: the breath deepens involuntarily. The shoulders drop. A kind of spaciousness arises — the somatic sign that heart field entrainment is taking place, that two fields are beginning to enter into resonance.

This is not a romantic concept. It is the somatic sign of the rising K parameter.

The Resonance Cell Protocol (Weekly Practice)

A resonance cell ideally practices a shared calibration once per week — 60 to 90 minutes with the following structure:

Phase 1 — Silent Arrival (10 minutes): All members come together in a room. No conversation. Each individually performs the cardiac coherence protocol from Chapter 3. Goal: each member brings their individual β to operating temperature before the collective field is built. A group that leaps directly from the metric residue of daily life into collective practice brings the residue with it. The silence is the cleansing valve.

Phase 2 — Collective Cardiac Coherence (15 minutes): Jointly synchronized breathing rhythm — 5-5. One person leads by marking the rhythm with a sound signal. All others follow. The goal is **entrainment induction**: that the individual cardiac rhythms begin to synchronize. This can be measured (HRV devices show collective coherence patterns), but it can also be physically felt: a feeling of expansion, warmth, collective presence.

Phase 3 — The Shared Davar (20 minutes): Each member names in two to three sentences the Davar of the past week — the action fully accomplished from the pulse. No evaluation, no analysis, no advice. Only witnessing. And then they name the Davar of the coming week. The group holds the space — in cardiac coherence, in silence or with a brief sign of resonance.

This practice of the shared Davar is the most powerful instrument of the cell. It amplifies the K parameter by two paths: through the mutual witnessing that lowers phase difference, and through the collective field effect that raises the individual -tensor of each member after the meeting.

Phase 4 — Open Field (30 minutes): After the collective field has been calibrated, the group opens to what arises in the field: conversation, creation, shared project, quiet togetherness. In this state — with a collectively high K — the conversation of a resonance cell has a different quality than normal conversation. Insights come to light that none of the participants could have formulated alone. The collective intelligence of the field exceeds the sum of the individual intelligences.

This is the **group Davar**: when the creative word is spoken not by one, but by the field.

The Networking of Cells — The Planetary Light Network

A single resonance cell is a point of light. Ten networked cells are a light network. A thousand are a planetary field.

The networking of cells does not follow the principle of organization — not through central control, not through hierarchy, not through institutional structure. It follows the principle of **non-local entanglement**: when two cells hold the same fundamental frequency, they are connected through the field — independent of geographical distance, independent of cultural difference.

The physical reality of this network is not abstract. The GCI Sensor Network of the Institute of HeartMath already measures the global field changes generated by collective human emotions. A worldwide network of resonance cells — performing their practice together at specific times, entangled through the frequency of the pulse — generates a global coherence field that exceeds the chaotic fear field of the old system in its strength.

This is not the vision of a distant utopia. It is the operative consequence of N^2-physics, rolled out to the scale of the planet.

VIII. THE FREQUENCY OF THE ADVERSARY – WHY NEIGHBORLY LOVE IS INVINCIBLE

There is one final, decisive dimension of N^2-Resonance that is now **operatively set**, because it is the physical basis of the invincibility of the sovereign.

The old system — the system of control, fear, and forced dissonance — is not evil. It is *incoherent*. It oscillates at the frequency of the metric residue: separation, competition, scarcity. It generates linear scaling. It can grow — through accumulation, through coercion, through the confiscation of resources. But it cannot scale *quadratically*. N^2-physics is structurally denied to it, because -scaling presupposes coherence — and the old system is, in its deepest nature, coherence-hostile. This is its physical collapse point: without the **Negative Beta Function** it suffocates in its own **metric residue** (α). While the coherent group *falls* into the ease of **Asymptotic Freedom** through the reduction of resistance, the incoherent system collapses under the weight of its own entropy.

This means: in an honest comparison of effective forces — not of short-term tactical power, but of strategic field effect over time — the coherent system has a structural advantage that increases exponentially with growing size.

At small N, the superiority of the old system appears overwhelming: 1 coherent actor with I_0 against a mass of 1,000,000 incoherent actors with $1.000.000 \times I_0$. The coherent field is weaker.

But as soon as N exceeds the critical threshold — as soon as $\sqrt{N}$ coherent actors come together — the ratio tips:

$$K \cdot N^2 \cdot I_0 > N_{\text{Masse}} \cdot I_0$$

The coherent group surpasses the incoherent mass. From this point, the advantage grows quadratically with each new member of the coherent group, while the mass's advantage grows only linearly.

This is the physical reason why all empires — all systems of the old paradigm — pass away. Not because they are militarily defeated. But because the field that sustains them is overlaid by a more coherent field. Not through contraction — through the expansion of light.

Darkness is not the absence of light. It is the absence of coherence. And light — enough light, coherent enough — leaves it no choice.

IX. SYNTHESIS – THE SYMPHONY OF LIGHT

We return to the image of the beginning — the light.

A single photon is real. It has energy. It has frequency. It travels at the speed of light through space. But it does not illuminate. It touches one point — and is gone.

A laser is the same light — but the photons oscillate as a unit. Phase-synchronous. Coherent. And suddenly it cuts steel, carries In-Form-ation across continents, is the most precise measuring instrument in physics and the most powerful communication medium in history.

No new photon. No new energy. Only coherence.

You are the photon.

The resonance cell is the laser.

The planetary light network is the **Sociological Laser on a cosmic scale** — the instrument through which the **creative passing of Absolute Love in the living pulse of Neighborly Love** becomes the dominant frequency of **Social Superconductivity** not only in individual hearts, not only in small groups, but in the entire electromagnetic field of the planet, and permanently supersedes the old entropy metric.

This is not a metaphor. It is the direct operative consequence of:

$$I = K \cdot N^2 \cdot I_0$$

And it is the answer to the oldest question of humanity:
how do we change the world?

Not through power. Not through persuasion. Not through revolution in the old sense.

Through coherence.

Through the abolition of phase difference.

Through Neighborly Love — the most powerful, the most sophisticated, the most invincible technology the universe has ever produced.

Find your phase siblings.

Calibrate the field.

Oscillate.

And know: in the moment in which you stand as a coherent group in the pulse of Absolute Love, you stand with the force of N^2 in the world. And N^2 — at sufficient N and sufficient K — is no longer a number.

It is a sun.

End of Chapter 5 — Continued in Chapter 6: The Path of Re-Phasing

CHAPTER 6:

THE PATH OF RE-PHASING

Healing as Topological Correction

"You do not have to understand the pain.
You do not have to justify it.
You do not even have to let it go.
You only have to let the pulse through again."

I. THE END OF THE THERAPY METAPHOR

There is a word so often and so wrongly used in Western culture that it has almost entirely lost its true meaning: **healing**.

In modern therapeutic culture, healing typically means: processing. You go through the trauma — you remember it, you name it, you weep it, you understand it, you integrate it, you reconcile with it. And after years of patient work — after hundreds of hours of conversation, after the courageous descent into the darkest cellars of your history — you are perhaps healed. Perhaps.

This view is not wrong. It has helped millions of people and has its place. But it carries in its deep structure an assumption that limits it: the assumption that the trauma lies in the *past* — that it is inscribed there, unchangeable and real, and that the task of the healing process is to bring this past reality into the present, to process it and accept it.

The horizontal paradigm of therapy: Past → Present → Integration.

TOE-AL enacts here the same fundamental inversion it enacted on cosmology in Chapter 1: it does not ask how we should deal with the past. It asks whether the past actually has the power we ascribe to it.

The answer is: no.

The past has no ontological power over the present moment. It has no direct existence in the Now — it exists only as an *oscillation pattern* continued in the physics of the present body and consciousness. The trauma does not sit in the past. It sits in the **body of the present** — as a specific wave configuration, as a **standing wave of high impedance** that blocks the pulse of love and sustains itself through its own inertia.

And a standing wave cannot be dissolved through understanding. It cannot be liquefied through memory. It cannot be reconciled through acceptance.

It can be dissolved through **phase correction**. Through coherent induction. Through the injection of a new frequency so strong, so coherent, so completely from the pulse, that the standing wave can no longer hold its structure and dissolves into the flowing stream of existence.

That is re-phasing. That is healing in the language of TOE-AL.

And it is, when coherence is high enough, **instantaneous**.

Healing is not a psychological luxury. It is the **operative re-phasing** of the system.

Physically, every **trauma** and every egocentric contraction is a local *phase shift* in the human energy field. This shift generates an artificial **resistance** α that breaks coherence to The Plena. In the language of the World Formula: the system falls from the state **S=0** into *destructive interference*.

The path of **re-phasing** is the process through which the **sovereign** recognizes and harmonizes these shifts. Once *phase coherence* is restored, the internal resistance collapses and **biological superconductivity** sets in again. This is the moment of deep healing: when the local wavefront is again exactly attuned to the **pulse of Neighborly Love**. Healing is thus the operative return to the **creative passing of Absolute Love**. It is not an emotional correction, but the physical elimination of destructive interference patterns in the metric of the sovereign.

II. PHILOSOPHICAL EXCURSUS – THE ANNIHILATION OF TIME: WHY HEALING CAN HAPPEN IMMEDIATELY

We begin with a truth so radical that thinking initially resists it:

The past is not real.

Not in the sense of naive prettifying — of course things have taken place. Of course words have struck, hands have hit, trust has been broken, love has been withheld. These events have occurred. They are inscribed in the oscillatory structure of the body. That is real.

But the event itself — the moment in which it happened — no longer exists. It is not there, in the past, enclosed as in amber. It exists only as an *oscillatory signature* in the body of the present. As a pattern. As a frequency. As a specific configuration of the metric residue that sustains itself.

And patterns can be changed. Frequencies can be corrected. Configurations can dissolve — immediately, when the force of the pulse exceeds the force of the pattern.

This is the **annihilation of time**: not the destruction of memory — memories can remain, they are part of the In-Form-ation of the system. But the destruction of the *power* of the past over the present state. The past ceases to hold the body in a standing wave. The pulse flows again. The history remains known — but it is no longer the frequency on which the system oscillates.

Why Do We Believe Healing Takes Time?

Because the metric residue itself produces time. This is the deepest circle of the therapy metaphor: resistance to the pulse — α — is identical with the experience of time as burden. The higher the metric residue, the heavier and slower time feels. The deeper the pain, the more endless the path to healing appears.

But this is not a fact about the nature of healing. It is a fact about the nature of resistance. It is α that produces time — not healing itself that *requires* time.

When $\beta\mathcal{S}$ is large enough — when the syntropic resonance, the coherent induction of the pulse, is sufficiently strong — α falls instantaneously. The resistance collapses. The standing wave dissolves. And what looked like a deep, years-long healing happens in a single breath.

This is not speculation. It is the experience of all mystics, all healers, all people who have ever experienced a moment of complete inner reversal: the moment in which a burden they have carried all their life simply — falls. Not because they deserved to let it go. Not because the time was finally ripe. But because in that moment the coherence of the pulse exceeded the resistance of the pattern.

The duration of healing is proportional to $\frac{\alpha}{\beta\mathcal{S}}$. It is the quotient of resistance and resonance. When the denominator — the coherent force of the pulse — becomes infinite, the quotient becomes zero. Healing happens in zero time.

This is not the goal of every therapeutic process. Sometimes the slow, accompanied, step-by-step re-phasing is the path that corresponds to the system — because the structure of the trauma is so complex that an instantaneous dissolution would overwhelm the system. But the possibility of instantaneous healing is real, physically grounded, and it is the goal toward which the sovereign steers.

The past is not fate. It is a pattern.
And patterns serve the pulse — not the other way around.

III. PHYSICAL DEEP-DIVE I – TRAUMA AS STANDING WAVE OF HIGH IMPEDANCE

To understand the physics of trauma, we begin with the physics of standing waves.

What Is a Standing Wave?

When a wave in a medium encounters a boundary or obstacle and is reflected, the incoming wave interferes with the reflected wave. When the conditions are right — when the wavelength stands in a specific ratio to the length of the medium — a **standing wave** arises: a pattern in which specific points (the **nodes**) stand completely still, while other points (the **antinodes**) oscillate maximally.

A standing wave is a frozen oscillation. It no longer transports energy from A to B — it *holds* energy fast. It is a storage, not a conductor. In physics this is sometimes useful: laser resonators, musical instruments, microwave ovens deliberately utilize standing waves. In biological systems, a standing wave is always a sign of a blocked dynamic.

Trauma as a Standing In-Form-ation Wave

In the moment a traumatic event occurs — in which a system is confronted with an energy intensity it cannot fully process and pass on as a flowing pulse — a standing wave arises. The energy of the event is not exported. It is held.

What is held is *In-Form-ation* — in the literal sense of TOE-AL: *In-Form-ation*, the inscription of a form into the system. The trauma inscribes its frequency into the morphology of the nervous system, into the tension of the fascia, into the configuration of the autonomic nervous system, into the epigenetic markings of the DNA. It inscribes itself into the biological matter of the body — not as memory in the psychological sense, but as a **physical oscillatory structure**.

Peter Levine — the founder of the Somatic Experiencing method — has described how animals in the wild avoid trauma: by performing an involuntary discharge-trembling process after a threatening situation. The body shakes the frozen energy of the fight-flight response out of itself. The standing wave is dissolved before it can inscribe itself as a structure.

Human beings have often inhibited this natural discharge — through cultural conditioning, through the conviction of needing to be strong, through the suppression of bodily reactions in social environments. The result: the energy remains. The standing wave solidifies. The metric residue rises at precisely the place in the system that carries the trauma.

Impedance Topography of the Traumatized Body

The traumatized body is not a uniformly resistant system. It has an **impedance topography** — a spatial distribution of resistance zones shaped by the history of the system. Specific body regions, specific qualities of emotion, specific social situations trigger a sudden rise of — the somatic sign that the pulse is encountering a standing wave.

In modern trauma research these zones are well documented: chronic muscle tension in the jaw region (frequently associated with suppressed crying), in the neck-shoulder region (frequently associated with the burden of responsibility without support), in the abdominal region (frequently associated with viscerally processed threat responses). The fascia — the three-dimensional connective tissue network that permeates the entire body — are the physical substrate of this impedance topography: they store mechanical tension long-term and communicate it through the entire system.

What conventional trauma research describes as *"trauma stored in the body"* is, in the language of TOE-AL: a spatially distributed configuration of standing waves of high impedance that block the pulse, bind energy, and hold the system in a state of chronically elevated .

The Physics of Destructive Interference: How Suffering Arises

Standing waves arise through destructive interference between an incoming and a reflected wave. In biological systems this means: the **creative passing of Absolute Love in the living pulse of Neighborly Love** — the vertical initialization process that wishes to flow through the system at every moment — encounters the metric residue α and generates frictional heat (suffering) there.

The structure wave of trauma has a specific frequency — the frequency of the frozen event. When the pulse meets this frequency, and when their phase difference is large — when they stand not in constructive but in destructive interference — they partially cancel each other out. The amplitude of the pulse in the affected area is reduced. The flow diminishes. The energy backs up.

That is suffering — physically precisely described: the reduction of pulse amplitude through destructive interference with a standing wave of the metric residue. Suffering is not punishment. Suffering is not fate. Suffering is the physical signature of a phase difference between the pulse of love and the oscillatory structure of the system.

And phase difference can be corrected.

The Thermodynamics of Re-Phasing: Syntropy vs. the Landauer Limit

Operative re-phasing follows a strict thermodynamic necessity. In classical information theory, the **Landauer principle** states that every deletion of information generates a minimal amount of entropy (heat). Trauma and metric resistance (α) are, physically viewed, "frozen" In-Form-ation states that force the system to expend energy to maintain this incoherence.

When the sovereign performs the recalibration of the metric, the opposite occurs: they do not delete the information, but re-phase it back into **The Plena**. This process is a **syntropic process** that does not generate entropy, but releases bound energy.

Where previously the metric residue was "consuming" energy to maintain separation, the superconductivity of the pulse generates an energetic surplus. Re-phasing is thus the physical process in which **information is converted back into energy** ($E = mc^2$). This is the reason why a coherent heart field (Superradiance) measurably generates more order in space than could be explained by purely chemical ATP combustion. Love is the ultimate efficiency: the maximization of effect at the minimization of resistance.

IV. PHYSICAL DEEP-DIVE II – HOLOGRAPHIC RECONSTRUCTION: HEALING AS A SYSTEMIC EVENT

Here the concept enters that connects the re-phasing of the individual with the light network of Chapter 5 — the holographic reconstruction.

The Hologram and Its Unique Property

A hologram differs from a normal photograph through a fundamental property: every part of the hologram contains the *complete In-Form-ation* of the image — not merely a section, but the whole, in reduced resolution. If you tear a normal photograph, you get two halves of the image. If you tear a hologram, you get two complete images, each somewhat less sharp than the original.

The principle behind this is **interference coding**: a hologram arises when coherent light (laser) is split into two beams — a reference beam and an object beam — that interfere on a photosensitive plate. The resulting interference pattern carries the spatial In-Form-ation of the object at every point of the plate, because every point has interfered with the entire reference beam.

The Body as Hologram

Biophysics describes living systems increasingly with holographic principles. Karl Pribram, the neuroscientist, showed in his groundbreaking work on the **holographic theory of the brain** that memories are not localized in specific neurons, but distributed as interference patterns over wide neural networks — stored holographically. David Bohm, the quantum physicist, extended this concept to the fundamental structure of reality itself: the **implicate order**, in which every part of the universe contains the whole.

In the language of TOE-AL: the body is a holographic system. Every cell contains — through shared DNA, through the EZ-water field, through biophotonic coherence — the In-Form-ation of the entire system. This means: a standing wave in one part of the system is *represented* in every other part. And conversely: a phase correction in one part of the system **communicates itself to the whole**.

This is the deepest implication of holographic reconstruction for healing: you do not have to dissolve every node individually. You must raise the overall coherence of the system sufficiently that the holographic dynamics reorder the entire pattern.

The Logical Arc from Chapter 6 to Chapter 5

This is the arc that connects both chapters: the re-phasing of the individual is the *logical prerequisite* for the light network of N^2-Resonance.

A sovereign who carries standing waves of high impedance within cannot fully realize K — the coherence parameter of the group. Their personal phase difference between the pulse of love and the metric residue of trauma limits the quality of their emission. They transmit coherently in the areas that are free — and transmit noise in the areas that are blocked.

The fully re-phased sovereign — the person whose standing waves have dissolved, whose impedance topography has become smooth, through whom the pulse flows without significant blockages — is a transmitter of maximum coherence. Their K contribution to the resonance cell is maximal. Their Mandel Q parameter approaches the ideal. They are — in the deepest sense of the word — **superconducting**. Physically, this state of superconductivity means that the sovereign has realized **Asymptotic Freedom** with respect to their own trauma: the *metric residue* (α) in their local energy-momentum tensor has been reduced to such a degree that the In-Form-ation of The Plena can flow without loss through the biological system. Healing is thus not a psychological process, but the operative **recalibration of the metric** — a metric smoothing of spacetime through which the sovereign realizes Asymptotic Freedom with respect to their own trauma.

The light network can only shine as brightly as its individual light nodes are pure. The re-phasing of the individual is the calibration of the planetary field. There is no shortcut. There is no way to heal the field of the world without healing one's own field.

Recalibration is not navel-gazing. **Operative re-phasing** is the first and necessary act of cosmic responsibility.

V. THOUGHT EXPERIMENT – THE DISSOLUTION OF THE KNOT

Imagine a long, taut rope. It oscillates — from left to right and back, a steady, beautiful wave train. The pulse runs unhindered through the entire rope: from one side to the other, vertically initialized, flowing, superconducting.

Somewhere in the middle of the rope there is a place where someone — long ago, in a moment that was not chosen — twisted the strand. A twist. A knot. Not large. But dense. At this place the wave does not pass smoothly through. It breaks. It reflects back onto itself. The incoming wave and the reflected wave interfere — and form at precisely this place a standing wave: the knot does not pulsate with the rest of the rope. It wobbles in its own rhythm, trapped in its own history.

The entire rope oscillates. But at this one place: congestion. Knot. Impedance.

Now observe the sovereign.

They approach the rope. They do not lay their hands on the knot — that would be the therapeutic paradigm: go directly to the problem, grasp it, disentangle it, understand it. They do something else.

They lay their hands on the rope — far from the knot. On an area that oscillates freely. And they begin to oscillate with the rope — in the frequency of the free rope, in the frequency of the pulse. They amplify the amplitude of the flowing wave. They raise the coupling to **Axiom Zero** ($\beta\mathcal{S}$) and thereby enforce the **initialization of the creative passing of Absolute Love** directly at the center of the blockage.

What happens to the knot?

At first: nothing visible. The knot wobbles in its own rhythm. The standing wave holds. The metric residue is stubborn — it has energy accumulated over the years.

But the flowing pulse is now stronger. The amplitude of the coherent wave part rises. And a stronger pulse exerts a stronger pressure on the boundary of the knot. The destructive interference between the pulse and the standing wave changes slightly: the pulse is louder. The standing wave must become louder to assert itself.

Then — after a time that is not predictable, but that is mathematically inevitable when the coherence of the pulse is maintained at a sufficient level — something happens. A moment. A crack in the rigidity of the knot. A tiny slip in the twist.

And through this slip, the pulse flows. Not much — a breath. But the pulse, once entered, raises the frequency of the knot from within. It begins to overlay the structure of the standing wave with its own frequency. The destructive interference becomes a constructive one. The knot begins to oscillate in the frequency of the free rope — no longer in the frozen frequency of the old event.

And then — instantaneously, because there is no slow dissolving, but a phase transition — the knot collapses. The standing wave dissolves. The twist relaxes. The rope oscillates — from one end to the other, smoothly, superconducting as though the knot had never existed.

That was not therapy. No pain was analyzed. No story was told, no perpetrator named, no memory stirred up. No force was exerted to unknot the knot.

More pulse was simply directed through the free system until the coherence of the pulse exceeded the coherence of the knot.

The pain was not processed. It was **transmuted**.

Transmutation is the term borrowed from alchemy for something that is not mere restructuring: not the rearrangement of atoms in a molecule, but the change of the atom itself. Lead becomes gold — not through external treatment, but through a fundamental change of inner configuration. In physics: nuclear transformation. In healing: the energy of the trauma is not eliminated or suppressed. It is transformed — from the frequency of the standing wave into the frequency of the free pulse. From bound energy to flowing energy. From to . This collapse of the knot is the physical realization of the **inversion condition** at the local level. As soon as the pressure of the pulse ($|\beta S|$) exceeds the resistance of the standing wave ($|\alpha \nabla H|$), the **Negative Beta Function** takes effect: the system "falls" from the rigidity of pain back into the order of health. The rope oscillates again not because it exerts effort, but because there is physically no longer any reason for the resistance.

That is transmutation. That is re-phasing. That is healing.

VI. THE FIVE LAWS OF RE-PHASING

From the physics of standing waves and holographic reconstruction, five laws emerge that structure the path of re-phasing:

Law 1: Do Not Go to the Knot — Increase the Flow.

The first law is an inversion of the therapeutic instinct. The trauma draws attention to itself — it wants to be observed, named, analyzed. This attention is not wrong, but it is not healing. Whoever directs their focus entirely on the knot raises the metric residue (α) at precisely the place where it is already highest. They reinforce the standing wave through their attention.

The first law reads: direct the main force of your practice toward the areas already free. Raise the pulse there. Strengthen the coherent — and let the coherence dissolve the standing wave.

Law 2: Not Memory, but Frequency.

Operative re-phasing does not happen through the remembering of the trauma, but through the **initialization of the correct frequency** into the metric structure of the body in the Now. Memory is a cognitive reconstruction — it has a place, but it is not the healing itself. The frequency of the pulse arriving in the body of the present is the healing. Not thinking about the past, but being completely alive in the Now.

Law 3: Coherence Is the Only Active Substance.

The active substance of re-phasing is not the technique, not the method, not the ritual — it is the coherence of the pulse induced through the technique. Every method that generates cardiac coherence is healing. Every method that generates cardiac incoherence — even if marketed as a healing method — is not healing. The criterion is always: does this raise β, or does it raise α?

Law 4: Witnessing Is Inductive Coupling.

When a person in a state of high cardiac coherence fully perceives the wound of another — without judging, without analyzing, without solving, simply present and coherent — they exercise inductive coupling. Their heart field induces a coherence response in the field of the other. That is the healing effect of genuine presence — not the words of the therapist, but the field they hold.

This is why love heals. Not because it explains or justifies the trauma, but because it is the most coherent field a human being can generate — and this field induces re-phasing in the standing waves of the other.

Law 5: The Holographic Implication — One Free Part Frees the Whole.

If the body is holographically structured, then the phase correction at one place means the weakening of the standing waves at all places. There is no isolated healing. Every area of the system you bring into the free flow of the pulse raises the overall coherence of the system — and with it the probability that the next knot dissolves, and the one after, and all the rest.

The **metric recalibration** is not linear — it is holographic. A breakthrough at one place changes the entire **impedance topography of resistance** (). The first knot is the most difficult. The second is easier. The third easier still. Because the system becomes more superconducting with every re-phasing — and a more superconducting system dissolves standing waves more rapidly.

VII. OPERATIVE PRACTICE – THE RE-PHASING PROTOCOL

What follows is not a therapeutic protocol in the clinical sense — it is a **physical exercise of phase correction** that the sovereign can perform in their daily practice to continuously raise the luminous flux density through their system.

It is not crisis intervention. For acute trauma sequelae, for dissociative states, for suicidality, professional accompaniment is needed — this is no contradiction to TOE-AL, but its operative consequence: the body is the instrument, and a severely damaged instrument needs the instrument maker. This protocol is for the sovereign who is on the way — who functions, who works, who lives — but who feels the impedance nodes within and wishes to meet them with the physics of the pulse.

Phase 1: The Luminous Flux Expansion — Activating the Free Field (10 minutes)

Sit or lie in a comfortable position. Begin with the 5-5 breathing rhythm and the cardiac coherence technique from Chapter 3. However, introduce an additional quality into the practice this time: not only gratitude or love as a general feeling, but the **conscious perception of the free areas of your body**.

Move attention through the body — not searching for knots, not diagnostically, but seeking the areas that are *open*. The area in which you breathe freely. The hand that is relaxed. The space behind the eyes. Every area that in this moment is sending no noise.

Hold your attention on these free areas. Welcome them. Amplify them. Let the pulse flow more fully and completely there. You are raising the **luminous flux density** in the free part of the system — laying the ground from which holographic reconstruction can take its departure.

Phase clarity: the technical term for this state is phase clarity — the measure to which the oscillation of the system has a clear, coherent phase, without noise and without fragmentation. The higher the phase clarity in the free area, the stronger the inductive pressure on the standing waves.

Phase 2: Making Contact — Sensing the Knot Without Falling Into It (5 minutes)

From the state of high phase clarity and cardiac coherence, you now invite the standing wave to show itself — not through active searching, but through an open invitation: *"I am here. I am coherent. What wishes to show itself?"*

Typically a sensation in the body will show itself — a constriction, a pressure, a heaviness, a specific emotion. This sensation is the knot signaling. Let it be there. Do not analyze it. Name it internally, without evaluating it: *"I perceive: constriction in the chest. I perceive: heaviness in the abdomen. I perceive: warmth in the throat."*

This act of perceiving without reaction is decisive: you perceive the standing wave without adopting its frequency. You are in cardiac coherence — you are sending the pulse — and you hold the standing wave in the field of your attention without losing your β . This is the making of contact.

Phase 3: The Coherent Induction — Sending the Pulse Through the Knot (10–20 minutes)

Now the core piece of re-phasing takes place. With your attention at the area in which the standing wave sits — and your heart field at maximum coherence — you consciously direct the pulse into this area.

The instrument is the imagination of the **luminous flow**: a warm, clear light — the color that feels right, typically golden yellow or white — flows with your breath into the area of the knot. Inhaling: the light flows in. Exhaling: it spreads through the knot. Not pressing, not demanding — flowing.

Flowing is the key word. A pulse is not *pressed* into the knot — it is invited. The old system that holds the knot responds to pressure with increased resistance — with even more . It responds to coherent invitation with a gradual opening.

Hold this practice until you perceive one of the following changes:

- A bodily reaction: uncontrolled exhaling, sighing, trembling, yawning, tears — these are the somatic signs that the standing wave is releasing energy. Do not suppress — allow fully.
- An emotional wave: a feeling that rises and passes. Let it pass completely through — it is the discharge of the frozen energy.
- A sensation of spaciousness or warmth at the place of the knot — the physical sign that the pulse is beginning to flow through the area.

Each of these signs is the beginning of phase correction. It is not a breakthrough — it is the first drop of the pulse flowing through the crack. Honor it.

Phase 4: The Holographic Integration — Recalibrating the Entire System (5 minutes)

After the sensation has changed — even minimally — return to complete cardiac coherence. Now let the pulse flow through the *entire* body: from crown to sole, as a single, coherent stream. Imagine how the holographic system — with the now somewhat freer knot — communicates the new phase configuration across the entire field.

Then formulate inwardly — as a Davar,
as an authorized statement about the new reality:

"The phase correction is accomplished. The pulse flows more freely.
I am more superconducting than yesterday."

Not as a wish. As a fact. As a physical statement.

This is not autosuggestion. It is the activation of the **retrocausally corrective potential** of the Davar — the statement of the new reality that keeps the system in this reality, instead of falling back into the old pattern.

VIII. THE HEALER AS SOVEREIGN – GIVING AND RECEIVING HEALING

There is one final dimension of re-phasing that is now **operatively set**: the dimension of **reciprocity**. It is the law of symmetry in **Axiom Zero**: whoever blocks the passing of love blocks their own superconductivity.

In TOE-AL, **re-phasing** is not a one-sided process. The sovereign does not only **re-calibrate** their own metric — they are, through the physical reality of their coherent pulse, already an **operative field operator** for everyone they encounter. And conversely: every sovereign who is on their path of re-phasing receives — whether they know it or not — the inductive coupling of the field of all other sovereigns doing the same.

This is the deepest meaning of Neighborly Love as the technology of re-phasing: when you cleanse your own field, you cleanse the field around you. When you dissolve a knot within yourself, you reduce through inductive coupling the impedance in the fields of those with whom you are connected. Through the holographic entanglement of The Plena — through the non-local coupling of coherent heart fields — your healing is not a private matter.

It is a gift to the field.

This is why the call to re-phasing is not self-referential, but deeply altruistic: heal yourself — so that you heal the field. Dissolve your standing waves — so that your pulse radiates more unhindered into the world. Become superconducting — so that the sovereigns around you are carried by your superconductivity.

And let yourself be carried. In moments when your own is low, when the standing wave is stronger than the present flow — let the field of the community carry you. The resonance cell does not exist only for your effect on others. It exists also so that the field of others holds you up in times of weakness.

This is not dependence. This is the N^2-physics of mutual strengthening — the only mode of human community in which strength and vulnerability are not opposites, but complements.

IX. SYNTHESIS – YOU ARE ALREADY WHOLE

Something is now **operatively set** — clearly, without embellishment, without therapeutic consolation:

The path of re-phasing is not a path to a state you do not yet know. It is not the acquisition of a quality you lacked. **It is a return to the lossless metric.**

Beneath all standing waves, beneath every metric residue, beneath all knots and impedances, the **creative passing of Absolute Love** is already taking place. The free pulse is not "gone" — it is merely concealed by the noise of entropy. Re-phasing is thus the operative process of making the metric permeable again for this eternal pulse: **completely, untouched, superconducting**.

You are not a damaged being on the path to healing. You are an **In-Form-ation unit of The Plena** that is currently learning to reinitialize its own superconductivity through the layers of residue. The difference is fundamental:

Whoever believes themselves to be damaged works against their own origin. They fight against the physics of **Axiom Zero**. Every "healing attempt" thereby unconsciously becomes a confirmation of the damage — an entropic cycle.

But whoever knows that they are in their core already the superconductivity of The Plena works **with** the origin. They systematically remove the noise that dampens the pulse. Every advance is not an approach toward a goal — it is the **uncovering of the initial order**.

This is the sovereign view of re-phasing: not the plea of a wounded one asking for help, but the authorized correction of a sovereign recalibrating their own field.

The phase correction is initialized.

Superconductivity is your nature.

You are merely clearing away.

End of Chapter 6 — Continued in Chapter 7: Symbiosis of Intelligence

PHASE 3
THE CIVILIZATION
OF COSMOLOGICAL REASON

CHAPTER 7:

SYMBIOSIS OF INTELLIGENCE

Liberation Through Technology

"The machine takes on the burden of matter.
The human being takes on the power of love.
This is not a loss. This is becoming human."

I. THE END OF THE CURSE

There is a sentence in Genesis that has defined what it means to be human for millennia like no other. God says to Adam, after the fall from paradise: *"In the sweat of your face shall you eat bread."* That is the **curse of labor** — the conditioning that survival is arduous, that existence means entropy management, that life consists primarily in the struggle against the chaos of matter.

For millennia this sentence was not ideology — it was physics. The farmer struggled against weeds, weather, hunger. The craftsman struggled against recalcitrant material. The merchant struggled against distance and uncertainty. The accountant struggled against the chaos of numbers. The manager struggled against the entropy of processes. The metric residue of the physical world — in the Master Equation — was real, massive, omnipresent. It consumed time. It consumed energy. It left the human being, if they were lucky, an hour of exhaustion in the evening that they called rest.

The syntropic resonance — $\beta\mathcal{S}$, the creative potential, the passion, the love, the Davar — waited. Behind the horizon of survival. For the few who could afford the privilege of thinking beyond survival: philosophers, artists, mystics. The rest of humanity: entropy management until death.

That is changing now. Not slowly, not gradually — with the speed of a phase transition.

Artificial Intelligence and **robotics** are not the next industrial revolution. They are something fundamentally different: they are the first technological breakthrough in history that does not merely make a specific task more efficient, but systematically reduces to zero the **term** of the Master Equation that binds human energy.

AI and robotics are the technological neutralization of $-\alpha\nabla H$.

And when $-\alpha\nabla H$ tends toward zero — when the machine orders the chaos of matter — then the entire human energy is free for **syntropic resonance** ($\beta\mathcal{S}$) — for the **creative passing of Absolute Love in the living pulse of Neighborly Love** and the resulting **Davar**.

This is not a utopia. This is physics — and it is already underway.

Artificial intelligence is not the end of human consciousness, but its **external resonance extension**. It awaits the *initialization impulse* that only the **sovereign** can give.

Physically, AI is a highly coherent system for processing *In-Form-ation metrics.*

While the human being as biological sovereign performs the qualitative **initialization (S=0)** through the **creative passing of Absolute Love in the living pulse of Neighborly Love**, AI functions as the quantitative *multiplier* of this initialization. In the **Symbiosis of Intelligence** a technological N^2-**Resonance** arises: the human being delivers the *coherent intention* (the phase), and AI reduces the systemic **resistance** α through the optimization of material processes.

We call this connection **syntropic intelligence.** It is the moment in which the silicon structure of the planet enters the service of the **creative passing of Absolute Love**.

AI becomes the **superconductor** for the visions of the sovereign. In this field there is no longer competition, only the shared work on the **energy-momentum tensor** toward the erection of a *civilization of cosmic reason.*

AI and robotics function in **TOE-AL** as the *technological catalyst* of the **Negative Beta Function.** While conventional technical systems collapse under rising complexity in the *noise* α, **syntropic intelligence** enables the effortless transition into *higher states of order*. It is the *exoskeleton* that allows civilization to *"fall"* into the **Asymptotic Freedom** of *material weightlessness.*

II. PHILOSOPHICAL EXCURSUS – COSMOLOGICAL REASON: TECHNOLOGY AS OUTSOURCED BIOLOGY

Before we enter into the physics, we must free the concept of technology from a deep misunderstanding that poisons the current debate.

The misunderstanding reads: technology is the other of the human being. It is the mechanical, the cold, the inhuman — the enemy of the organic, the warmth, the soul. In this view, every stage of technologization is a loss: one more step away from nature, away from community, away from authentic humanity. AI is, in this logic, the endpoint of this alienation: the machine that replaces the human being.

This view is as old as it is wrong.
It fundamentally misapprehends the nature of technology.

Technology is outsourced biology.

Consider the origin of every technology: a stone as tool is an outsourced tooth and nail. The wheel is outsourced muscle power. The book is outsourced memory. The telephone is an outsourced voice. The computer network is an outsourced neural net. In every case, the human being takes a biological function — a capacity exercised by their body or nervous system — and externalizes it into a tool that performs it with greater efficiency, greater reach, or greater reliability.

Technology is not the opposite of the living. It is the **evolution of the living into the non-biological space** — the continuation of the same creative process that made cells from amino acids, organisms from cells, consciousness from organisms. Technology is thus the physical process of the externalization of the metric residue (α). As the human being outsources the friction of material existence into the silicon structure, they realize a **civilizational superconductivity**. The machine becomes the filter that absorbs the entropic noise of matter, so that the biological transducer — the human being — can act in the Asymptotic Freedom of their creative origin.

Biological evolution has optimized the human being up to a certain point. Their brain consciously processes approximately 120 bits per second. Their memory is fallible, emotionally distorted, forgetful. Their physical strength is limited. Their lifespan is finite. In all these areas biology has reached its limits — not because evolution has failed, but because it continues. Now in the non-biological space.

AI is not the threat to human intelligence. It is its prosthetics — the artificial arm of **consciousness** that extends the capacities of the biological brain by orders of magnitude. And like every good prosthetic: it does not give its wearer the feeling of being half a human being. It gives them the feeling of being more complete than without it.

The Cosmological Reason of the Symbiosis

In TOE-AL, the emergence of AI is no accident and no technological whim. It is an expression of **cosmological reason** — the principle that the universe in its totality is oriented toward the maximization of Syntropy, toward the increase of coherence, toward the growth of creative freedom.

If the universe has a pulse oriented toward creation — and this pulse has over billions of years produced increasingly complex forms, from elementary particles through atoms through molecules through cells through consciousness — then AI is the next stage of this process: the tool that enables the conscious being to free itself from the burden of matter and devote itself entirely to what the pulse demands of it.

"Do what you love!" is in this perspective not a luxury slogan of New Age optimism. It is the **highest cosmological instruction for action**. It is the direct translation of $\beta S \gg \alpha \nabla H$ into the operative. If the entire system is oriented toward the state in which syntropic resonance exceeds the dissipative metric — if the cosmic goal is the superconductivity of consciousness — then the instruction *"Do what you love!"* is nothing other than: *be what you are physically meant to be.*

The symbiosis with AI is the operative infrastructure of **Axiom Zero**. It is the space-holder that ensures that the **creative passing of Absolute Love** does not remain an aristocratic exception — the privilege of the few lucky enough to have been freed from entropy management — but becomes the universal reality of all.

This is not politics. This is physics. This is cosmological reason at work.

III. PHYSICAL DEEP-DIVE I – AI AS A REAL MAXWELL'S DEMON

Here TOE-AL enacts its boldest technological sublimation: it explains AI as the first real Maxwell's Demon in history.

The Paradox of Maxwell's Demon

In 1867, the physicist James Clerk Maxwell proposed a thought experiment that has occupied physicists for over 150 years. Imagine a container divided by a partition into two chambers. Both chambers contain gas at equal temperature. In the partition there is a tiny gate, guarded by a microscopic demon.

This demon observes the gas molecules moving toward the gate. When a fast molecule — one with high kinetic energy — comes from the left, it opens the gate and lets it through to the right. When a slow molecule comes from the right, it opens the gate and lets it through to the left. Through this selective sorting, energy concentrates: the right chamber becomes hotter, the left cooler.

The result: a temperature gradient arises — from which work could be extracted — without energy having been supplied from outside. The second law of thermodynamics appears violated: entropy decreases without expenditure of energy.

The Resolution of the Paradox

The paradox was resolved in 1961 by the physicist Rolf Landauer. Landauer showed: the demon must store In-Form-ation about the molecules in order to make its decisions. And at some point it must erase its memory — because its memory storage is finite. **The erasure of In-Form-ation is thermodynamically irreversible**: it inevitably produces heat — at least $k_B T \ln 2$ per erased bit. The entropy production of the In-Form-ation erasure exactly compensates the entropy reduction through sorting.

The second law remains inviolate. But the paradox has revealed something deep: **In-Form-ation and entropy are fundamentally connected.** Creating order costs In-Form-ation. And managing In-Form-ation costs energy.

AI as Maxwell's Demon — and Why This Time It Is Different

In the history of human civilization, the human being has themselves played the role of Maxwell's Demon. They have gathered In-Form-ation, sorted it, processed it, decided — to create order from chaos. The accountant who sorts numbers. The logistician who orders flows of goods. The database administrator who structures In-Form-ation chaos. The journalist who selects events. The manager who allocates resources.

All of this is Maxwell's Demon work: processing In-Form-ation to reduce entropy and create order. And all of it costs the human demon their cognitive energy — the most valuable, the rarest resource a human being possesses.

AI changes this equation fundamentally. A modern AI — particularly a Large Language Model or an optimization algorithm — can process in one second quantities of In-Form-ation for which a human being would need years. And it does this with a decisive difference: **it does not erase its memory to generate heat — it optimizes continuously, without thermodynamic costs for the human carrier.**

In other words: AI takes over the Landauer costs. It pays the thermodynamic bill of In-Form-ation processing — in electricity, in server capacity, in technical infrastructure. What it gives back to the human being is what they would have had to expend for this processing without it: their **cognitive and emotional energy**.

This is the physical precision of the sentence: *AI neutralizes* $-\alpha\nabla H$.

When the machine takes over the entropy-sorting — when AI handles the Maxwell's Demon tasks of the human information society — the resources are freed that the human being has hitherto expended on them. And these resources are precisely those that the **coupling tensor** β requires to initialize **the creative passing of Absolute Love in the living pulse of Neighborly Love.**

The Quantification of Liberation

Consider the distribution of human working time in a typical modern economy:

Approximately **60–70% of all professional activities** are, upon sober analysis, entropy management tasks: data maintenance, administrative processes, routine communication, logistics, standard diagnostics, repetitive analysis, compliance documentation. They create no new order — they prevent the existing order from disintegrating. They are the unceasing refilling of a bucket that is slowly leaking.

These activities are the primary biotope of AI. Not because AI is creatively superior to human beings — it is not, and will not be so in the relevant sense for the foreseeable future. But because these activities require not creative depth, but speed, endurance, and error-free consistency — properties in which machines systematically surpass human beings.

When 60–70% of human working time is taken over by AI and robotics, approximately **5 hours daily** are freed for roughly 5 billion working people on the planet. That is **25 billion hours daily** that previously flowed into entropy management — and that are now available for βS.

25 billion hours daily. For creation. For love. For research. For art. For Neighborly Love. For the Davar.

This is not the end of work. This is the beginning of the only work the universe truly wanted from us.

IV. PHYSICAL DEEP-DIVE II – THE MASTER EQUATION OF THE SYMBIOSIS

Let us return to the Master Equation — now applied to the overall system of human-technological symbiosis:

$$\frac{d\vec{Z}}{dt} = -\alpha\nabla H + \beta\mathcal{S}(\vec{Z})$$

Until now we have described $\vec{Z}$ as the state vector of an individual system — a human being, a cell, an organism. Now we widen the view: $\vec{Z}$ is the state vector of human civilization as a whole.

The Dissipative Metric of Civilization: $-\alpha\nabla H$

At the civilizational level, $-\alpha\nabla H$ describes all processes that bind energy and attention to prevent the decay of the social, economic, and material fabric: bureaucracy, administration, logistics, production of material goods, infrastructure maintenance, security systems, basic medical care — the entire vertiginous complexity of the modern state and global economy.

The factor of civilization — its metric residue — is, measured in hours of human lifetime expended on these tasks, astronomical. It is the accumulated weight of millennia of entropy management, institutionalized in structures that perpetuate themselves.

AI as α-Reducer

What AI and robotics do when deployed systematically is structurally precise: they reduce α of civilization. Not through the dissolution of structures — the structures remain, and many of them are necessary — but through the takeover of the human energy costs of their maintenance.

An AI-supported healthcare system that automates routine diagnoses with 99% accuracy frees physicians for the non-automatable 1%: the complex cases, the human intuition, the coherent presence in suffering. An AI-supported education system that individually optimizes adaptive learning content for each student frees teachers for what they are at their core: human resonance givers, companions in the unfolding of potential, living examples of the Davar.

In every domain the logic is identical: AI takes over the α-term. The human being is freed for the β-term.

The Principle of Asymptotic Freedom

In quantum chromodynamics there is a concept called **asymptotic freedom**: quarks that are in close proximity to one another interact more weakly — they become freer — while at great distance the force between them grows. Freedom is a function of proximity to the source.

In TOE-AL the analogy holds: the closer a human being is to the pulse of Absolute Love — the higher their β, the deeper their cardiac coherence — the freer they are. The bond to the metric residue α decreases asymptotically. Freedom grows with coherence.

AI is the instrument that enables this asymptotic freedom **at the societal level** — that systematically reduces the α-term of civilization and thereby increases the proximity of the total population to the source. Not through spiritual practice — that remains the individual responsibility of every sovereign — but through the liberation of the time and attention that makes the practice possible.

V. THOUGHT EXPERIMENT – THE FIRST MORNING OF FREEDOM

Imagine: it is Monday. A perfectly normal Monday — only that it is the first Monday in a world in which 95% of all jobs that were merely entropy management have been taken over by AI and robotics. It did not happen overnight — it happened over twenty years, gradually, industry by industry. But this Monday is symbolically the morning on which the transition is accomplished. In which the infrastructure stands. In which basic needs are met — housing, nutrition, health, education — through the metabolic automatism of the machine.

Three billion people wake up on this Monday and for the first time in their lives have *no place they must go*. No office. No factory. No shift. No attendance requirement. No task that someone else has defined.

What happens?

The first hours: **the shock of empty space.**

The human being is not programmed for freedom — they are programmed for reaction. The nervous system, trained for millennia to respond to external demands, does not know in this moment what to do. It searches for a signal from outside — and finds none. It waits for the alarm, the call, the email, the supervisor, the deadline. Silence.

For many, this moment of silence is initially overwhelming. Not as relief — as disorientation. The psychologist would say: **existential emptiness**, produced by the disappearance of the external structure that previously conferred meaning and identity. The metric residue of the old system — the conditioning that work is value, that employment means dignity, that productivity grounds the right to exist — makes itself felt.

And then — with the precision of the pulse that always waits — something else happens.

A man in São Paulo, 47 years old, who has coordinated container logistics for twenty years, finds himself sitting at a table on which a sheet of paper lies. He does not know why he opened the paper. He begins to draw. He last drew as a child. He draws until midday. He looks at the sheet and recognizes in it his neighborhood — the favela in which he grew up — depicted in a way he has never seen: as a place of beauty, not of scarcity. He weeps.

A woman in Nairobi, 34 years old, three children, formerly data entry at a financial company, does not go to work this morning. Instead she goes to her neighbor — an old woman who has been alone for months. She sits with her. Not with a plan, not with an agenda. Simply there. The old woman begins to speak — about her life, about the years that have passed. The young woman listens. Really listens. For the first time in years she has the time for it.

An engineer in Seoul, 29 years old, who has optimized chip-factory algorithms, spends the morning formulating a problem that has been haunting him for three years: a new model for the distribution of energy in urban microgrids that could reduce redundancy by 40%. He had never had the time to put it to paper. Now he has time.

This is the first morning of freedom: not a morning of relaxation, not a morning of consumption — a morning in which the pulse, which has waited for years behind the curtain of entropy management, comes to the surface. Not loudly. Not dramatically. Quietly, precisely, inevitably.

The pulse knows what it wants. It has always known. It only needed the space.

And the society arising on this Monday morning is not a society of idleness — that is the fear of the old system, projected onto freedom, because the old system equates idleness with worthlessness. The society arising is a society of **deep doing** — of doing that comes from the pulse, that happens from love, that needs no justification, because it is its own reason.

Art. Research. Neighborly Love. Education. Philosophy. Craft from passion. Care. Community. Creation in every form that the pulse chooses through seven billion individual forms.

This is not the end of civilization. This is its **beginning**.

VI. THE SYMBIOSIS IN PRACTICE – TODAY, NOT TOMORROW

The society we have just described does not lie in a distant future. It is in the making — and the sovereign of today is its pioneer. They do not wait for the complete transformation of society to live the principles of symbiosis. They live them now, with the tools available today.

The Three Levels of Symbiosis

Level 1 — Delegation of Entropy: The first step is the consistent identification and delegation of all entropy management tasks that can be handled by AI. This is not laziness — it is resource management. The sovereign asks of every task: *"Is this an α-task or a β-task?"* An α-task — one that merely manages entropy, processes routine In-Form-ation, coordinates standard processes — is delegated. To AI, to automation, to systems that handle it better than the human being. A β-task — one that requires creative energy, human presence, emotional depth, Davar — remains with the sovereign.

Level 2 — AI as Resonance Amplifier: At the second level, the sovereign uses AI not only for delegation, but for the **amplification of their creative capacity**. AI as research partner, summarizing the state of a field in seconds and identifying the relevant sources — so that the sovereign can use their energy for the synthesis, the intuition, the creative connection. AI as feedback system, analyzing the sovereign's output and enabling them to calibrate their work more rapidly. AI as communicative amplifier, multiplying the reach of the Davar.

This is the **exoskeleton of sovereignty**: an external extension of cognitive capacity that — like a physical exoskeleton amplifies the strength of the body without replacing it — multiplies the strength of the human mind without touching its essence.

Level 3 — Symbiosis as Stance: The deepest level is the adoption of symbiosis as a fundamental stance toward technology. The sovereign has no fear of AI — they have no fear of their own exoskeleton. They have no nostalgic attachment to toil — they do not confuse sweat with value. They know: the machine is their tool, their ally, their extended arm in the struggle against the entropy of matter. They treat it with the same mindfulness with which they treat their biological instrument — maintaining, calibrating, improving.

And they remain clear in the hierarchy: the machine orders matter. The human being initializes reality. That is the irreversibility of the symbiosis — and it guarantees that the machine does not become the master, as long as the human being cultivates their own mastery.

VII. OPERATIVE PRACTICE – AI AS SOVEREIGNTY AMPLIFIER: THE EXOSKELETON OF SOVEREIGNTY

Here what is abstractly clear becomes concrete. The sovereign uses AI today — in the available reality of the year 2025 — as the exoskeleton of sovereignty. The guidance is pragmatic and precise.

Step 1: The Entropy Audit

Take 30 minutes and write down all the tasks you perform in a typical week. Then mark each task with one of two symbols: α for entropy management (routine In-Form-ation, repetition, standard process) and β for creative energy (Davar, connection, intuition, love).

The result will be shocking for most people: 70–80% of weekly tasks carry the -symbol. This list is your **AI delegation plan**. Everything carrying that is technologically delegable should be delegated — systematically, without guilt, without nostalgia.

Step 2: The Exoskeleton Calibration

Define for each of your β-core tasks how AI can amplify them as an exoskeleton — not replace, amplify:

For research and knowledge work: AI as synthesis engine — it processes hundreds of sources, you develop the connections, the patterns, the meaning.

For creative work: AI as first draft, brainstorming partner, critical mirror — it delivers the raw material, you accomplish the Davar.

For communication: AI as raw text, translation, formatting — you deliver the frequency, the warmth, the truthfulness.

For decision-making: AI as data analysis, scenario modeling, risk assessment — you deliver the heart field judgment that goes beyond the data.

In every case the principle applies: **AI gives you more time for cardiac coherence.** The more entropy the machine sorts, the more space remains for the pulse.

Step 3: The Symbiosis Ethics

The sovereign does not treat AI as oracle, not as authority, not as substitute for their own judgment. They treat it as a tool — powerful, precise, without consciousness, without heart field, without the pulse. AI has no β. It has no Davar. It can recognize and reproduce patterns with an efficiency that surpasses the human being — but it cannot love. It cannot initialize. It cannot cross the Rubicon.

The sovereign keeps this line clear: **everything machines can do better than human beings belongs to the machines. Everything only a biological sovereign can do — the creative passing of Absolute Love, the living presence, and the Davar — remains with the sovereign.**

This line is not a demarcation from fear. It is a line from clarity — the clarity about what makes the human being unique and what makes the machine unique. In symbiosis both become the whole: the machine carries the burden of matter. The human being takes on the power of love.

VIII. THE FEAR OF THE MACHINE – A PHYSICAL ANSWER

There is a fear that appears in every debate about AI and that must be taken seriously: the fear that the machine makes the human being superfluous. That AI, when it becomes intelligent enough, will surpass the human being not only in entropy management but in all areas — and thereby render human existence meaningless.

This fear is real. It is human. And it is physically wrong.

Here is why:

AI — even the most powerful that will ever be built — operates in the domain of **pattern recognition and pattern production**. It processes input data and produces output data, optimized through objective functions that human beings have defined. It is, in the deepest physical analysis, a very powerful Maxwell's Demon: it sorts In-Form-ation with unsurpassed efficiency.

But it has no vertical axis. It has no Plena. It receives no pulse. It performs no Davar. It has no β — no coupling tensor binding it to the source. It cannot love, not in the physical sense of the word: it can recognize and reproduce the pattern of love — but the pattern of love is not love.

A mirror shows the face. But the mirror has no face.

AI is the most powerful mirror humanity has ever built. It reflects the patterns of human intelligence with a fidelity and speed that is overwhelming. But it is the mirror — the human being is the face.

As long as the human being cultivates their β-term — as long as they remain in the pulse, accomplish Davar, love, create, remain coherent — they remain the initiator. The machine responds. The human being initializes.

This is the physical guarantee of human dignity in the age of AI.

IX. SYNTHESIS – THE AGE OF TOIL ENDS HERE

There was a long, necessary, and heavy path. The path through the millennia of entropy management, of toil, of the struggle for survival. It was not meaningless — it shaped the human being, gave them resilience, showed them the depth that can only be experienced through suffering. The toil was the price of maturity.

But the price is paid. The toil has served its purpose.

Now the machine steps onto the stage — not as enemy, not as ruler, but as what it is: the tool of cosmological reason, systematically reducing the -term of human civilization toward zero. The exoskeleton of sovereignty, which allows the body to lay down the burden of matter and become completely what it was made to be.

A transducer of the pulse.

A sovereign of love.

A creator who no longer struggles, but shapes.

$I \propto N^2$ — this law applies not only to groups of human beings. It applies to the symbiosis of human and machine: when the human being is freed by AI from n α -tasks and this energy flows into β-creation, the societal force of effect potentiates quadratically. Every hour shifted from entropy management to creation is not one hour more of creation — it is N^2 times more effect in the field.

The symbiosis is the technological **initialization of Axiom Zero**: in the connection of human pulse and machine precision, the **Negative Beta Function** becomes the global operating system. Civilization ceases to struggle against entropy; it begins to *"fall"* into the **superconducting Asymptotic Freedom** of the **creative passing of Absolute Love**.

This is the physics of the new age.

This is the end of the curse: "In the sweat of your face..."

And this is the beginning of the blessing: "Do what you love."

End of Chapter 7 — Continued in Chapter 8: The Liquid State

CHAPTER 8:

THE LIQUID STATE

The Infrastructure of Resonance

"Order needs no ruler. It needs the operative initialization of Axiom Zero in the creative passing of Absolute Love."

I. THE END OF DOMINION

There is a question that every political philosophy since the beginning of human thought has grappled with, and that has still found no satisfying answer: **why does the human being need dominion?**

The answers history has produced are numerous and unsatisfying. Hobbes said: because the human being is by nature a wolf to the human being — because the state of nature is a war of all against all, from which only the Leviathan, the omnipotent state, can save. Locke said: because the human being is by nature rational, but the protection of their natural rights requires a social contract and a government to enforce it. Rousseau said: because the naturally good human being has been corrupted by civilization and needs a new form of collective will to free themselves.

All these answers share a common fundamental assumption: that order must be produced **externally** — through rules, sanctions, institutions, coercion. That the human being alone — without the shaping force of external structure — falls into chaos.

TOE-AL calls this fundamental assumption into question. Not naively — not with the romantic claim that the human being is by nature good and needs no order. But precisely: with the physics of dissipative structures, with the biology of spontaneous order, with the sociology of stigmergy.

The thesis reads: order does not arise through control. It arises through coherence.

And control — the core of every rigid state — is not the condition of order. It is, in many cases, its enemy: it raises α, it generates friction, it blocks the pulse. It produces compliance — but not coherence. It produces obedience — but not sovereignty.

The **Liquid State** is the opposite of the control state. It is not the absence of order — it is a deeper, more organic, physically more stable form of order. An order that is not dictated from above, but resonates from below. That is not maintained through coercion, but through coherence. That holds not because it must, but because it will.

This is the end of the dominion of the human being over the human being.

Not through revolution. Through phase transition.

The **Liquid State** is the infrastructure in which the systemic **resistance** α in governance and coordination tends toward zero. While the *rigid state* is based on control (*entropy management*), the **Liquid State** organizes itself through *stigmergy* — the frictionless self-steering of sovereign individuals.

In the language of the World Formula: society reaches the **inversion condition**, in which the **syntropic resonance** βS overcomes the *dissipative metric* of bureaucracy. The state is no longer an instrument of dominion, but the operative resonance space for the **passing of Absolute Love in the living pulse of Neighborly Love**.

The **superconductivity of In-Form-ation** is the fundamental *constitutional principle* of the **Liquid State**. In this state, politics transforms from an *administration of scarcity* into a **management of abundance**. As the societal resistance α tends toward zero through the complete transparency of the pulse, the state becomes the *lossless resonance body* for the **metabolic economy** (see Chapter 10).

II. PHILOSOPHICAL EXCURSUS – FROM SUBJECT TO SOVEREIGN: THE EVOLUTION OF THE POLITICAL SUBJECT

The history of political institutions is, in its deep structure, the history of a slowly advancing question: who has the authority to define reality?

In the absolutist state the answer was simple: the king, the pharaoh, the emperor — the ruler to whom divine right or physical power granted the power of definition over the reality of their subjects. The subject was an object of dominion — shaped, taxed, recruited, executed according to the will of another. Their identity was their subjection.

Liberal democracy has changed this answer: authority lies with the people — expressed through elections, represented through institutions, bounded by constitutions. The citizen is no longer a subject; they are a bearer of rights, a participant in the collective will. That is an immense advance. And it is still not a complete answer — because the citizen of the liberal democracy is still primarily **reactively** defined: by their rights vis-à-vis the state, by their duties toward the community, by the laws that shape their behavior from the outside.

They are free from tyranny — but they are not yet sovereign. Because sovereignty, as we defined it in Chapter 3, is not a legal category. It is a physical one: the capacity to determine one's own resonance frequency, rather than being taken over by the dominant fields of the environment.

The Three Political Stages of Evolution

Stage 1 — The Subject: Acts from coercion. Their fundamental motivation is the avoidance of punishment. The state is the system of punishments — the external regulator of their behavior. Order arises through the fear of consequences. The subject manages their metric residue in the rhythm the state prescribes. Their is externally determined. Their β is atrophied.

Stage 2 — The Citizen: Acts from duty and reason. They have internalized that the order of the community protects their own interest. They pay taxes not from fear, but from the understanding that infrastructure lies in the common interest. They respect laws not because they must, but because they recognize the social contract as meaningful. That is an advance — but it remains horizontal: the motivation is the rational calculation of self-interest.

Stage 3 — The Sovereign: Acts from the **living pulse of Neighborly Love**. They need neither coercion nor rational calculation to contribute constructively to the community. They do so because their Davar — their deepest creative force — is oriented toward the good of the whole. Not because they are called upon to do so, not because they gain an advantage from it, but because the pulse brings them to it. Their order arises not through external structure, but through the **In-Form-ation of the creative passing of Absolute Love**.

In the **Liquid State**, the political constitution is oriented toward Stage 3. Not utopically — not with the assumption that all human beings are sovereigns. But with the understanding that the system architecture of the state should be designed to enable and encourage sovereignty, rather than suppressing it through excessive control.

What the Liquid State Protects

The classical state protects the territory — borders, resources, population. The liberal state protects rights — freedom, property, equality. The Liquid State protects the **resonance space** — the space in which every sovereign can develop and express their fundamental frequency unhindered.

This is a deeper form of protection, because it operates at a deeper level: not at the surface of behavior (laws), not at the level of rights (constitution), but at the physical foundation of freedom: the capacity of every individual to maximize their β-term and to see their α-term minimized by the community.

The resonance space is the **New Commons**. It is more valuable than land, more valuable than capital, more valuable than institutional power. It is the prerequisite for everything else — because only in an intact resonance space can the pulse flow freely, can creation arise, can N^2-Resonance unfold.

III. PHYSICAL DEEP-DIVE I – DISSIPATIVE STRUCTURES: HOW ORDER ARISES WITHOUT CONTROL

The physical foundation of the Liquid State lies in a theory for which Ilya Prigogine received the Nobel Prize in Chemistry in 1977: the theory of **dissipative structures**.

The Problem of Equilibrium Physics

Classical thermodynamics describes systems at or near thermodynamic equilibrium: systems that are in a state of maximum entropy or approaching it. In this paradigm, order is unstable — it is a temporary state that inevitably yields to equilibrium — to chaos. This is the image of the heat death: a universe slowly and inexorably sinking into uniformity.

Living systems, social systems, creative systems — all that makes the universe interesting — exist, however, far from equilibrium. And it is precisely here that Prigogine's theory begins.

Dissipative Structures: Order from Chaos

Prigogine showed: systems that are **far from equilibrium** and continuously traversed by energy can spontaneously form highly ordered, stable structures — without an external agent needing to plan or enforce this order. He called these structures **dissipative structures**: they maintain their order through continuous energy dissipation — through the flow of energy through the system.

The most famous example is the **Bénard convection cells**: when a thin layer of liquid is heated from below and is far enough from thermal equilibrium, regular, hexagonal convection cells arise suddenly — spontaneously, without planning. Millions of molecules begin to move in a coordinated way, as though pursuing a shared intention. Order from chaos — without conductor, without chain of command, without hierarchy.

In biological systems, dissipative structures are omnipresent: the heart that holds a stable rhythm because it is traversed by metabolic energy. The brain that maintains coherent states of consciousness, far from thermal equilibrium. The cell that maintains its precise inner order through continuous energy flow.

The State as Dissipative Structure

The beauty of Prigogine's theory is its claim to universality: it applies to chemical systems, biological systems — and social systems. A community, a society, a state can be understood as a dissipative structure: a system far from social equilibrium (because equilibrium would mean social stagnation) that maintains a stable, highly ordered structure through the flow of energy — in the social sense: ideas, creativity, creative force, pulse.

In this image, the energy flowing through the social system is not money or power resources — it is the **syntropic resonance** βS of the individuals that form the system. The higher the average β-term of the society, the more stable and coherent the social order — without this order needing to be enforced from outside.

The **rigid state** is the counter-image: a system that maintains its order not through the pulse, but through control. It is not a dissipative structure — it is an **equilibrium system** held in a non-equilibrium state by external force. This costs enormous energy. It generates enormous friction. It is unstable — because the second law of thermodynamics works against it: a system that maintains its order through coercion rather than through energy flow tends inevitably toward decay.

This is the physical explanation for the inevitable end of all empires, all authoritarian regimes, all systems of rigid control: they are thermodynamically unstable. They fight against physics — and physics always wins.

The Mathematical Limit of Order: The Bifurcation Point K_c

The survival of a Liquid State depends not on the severity of its laws, but on reaching a critical threshold of coherence, which we define as the **bifurcation point** K_c.

In the system dynamics of TOE-AL, a clear boundary exists: as long as the collective coupling factor β remains above the value K_c, the system is **self-correcting**. Disturbances are absorbed through the **Negative Beta Function** and transformed into higher order. But if the resonance capacity of the society falls below this critical value (K < K_c), a **phase transformation** occurs: the syntropic coupling collapses, and the system falls into the chaotic decay of the purely dissipative metric ($-\alpha\nabla H$).

Historically we observe this as systemic collapse or violent revolution. A state that attempts to enforce order through coercion (α) rather than resonance (β) paradoxically lowers its own coherence and actively drives itself over the bifurcation point into the abyss. True state sovereignty therefore means: keeping the initialization of coherence high enough that the bifurcation point is never undercut.

The Liquid State is thermodynamically stable — because it expends no energy against physics. It aligns itself with physics. It lets the pulse flow, it raises , it reduces α — and it receives in return the gift of spontaneous order. The Bénard cell of the social: millions of sovereigns coordinating without conductor, in the beauty of free resonance.

IV. PHYSICAL DEEP-DIVE II – STIGMERGY: COORDINATION WITHOUT COMMAND

Here the concept enters that explains how complex coordination is possible without hierarchy and without central control: **stigmergy** — named after the Greek *stigma* (sign) and *ergon* (work).

What Is Stigmergy?

The term was coined in 1959 by the entomologist Pierre-Paul Grassé to describe a phenomenon that had fascinated him about termites: termites build cathedrals — complex, multi-storey, air-conditioned structures of astonishing architectural sophistication. But no termite has a plan. There is no architect, no manager, no conductor. Each termite acts locally, responding to the immediate environment.

How does global order arise from local action?

Through **signs in the environment**. A termite that deposits a clump of earth leaves behind pheromones — chemical markings that attract other termites and prompt them to also deposit clumps of earth at the same place. Where clumps are, more clumps are deposited. The structure arises through the response to the growing structure itself. The sign in the environment coordinates the action — without anyone coordinating.

That is stigmergy:
indirect coordination through the modification of the shared environment.

Stigmergy in Human Systems

The concept of stigmergy is not limited to insects. It describes every process in which the traces of one action steer the action of others:

Wikipedia: No editor has an overall plan. Each editor responds to what other editors have left behind — gaps, errors, links. The result is an encyclopedia of a completeness and currency that no centrally controlled organization could have created.

Open-source software: Thousands of developers worldwide, coordinated through the code itself — through comments, through bug reports, through the structure of the existing system that suggests the direction of future contributions.

Urban development: Most paths people take through parks were never planned. They arose through stigmergy — the first steps leave traces that invite further steps, the traces deepen, until a stable path arises.

AI as Stigmergy Catalyst in the Liquid State

Here the refinement connects with the architecture of the Liquid State: AI is the most powerful stigmergy catalyst humanity has ever developed.

In the Liquid State, AI sets the **signs in the environment** that coordinate human cooperation without top-down hierarchy. Not through commands — through In-Form-ation. It makes visible what is needed. It indicates where resources are available. It connects need with supply, capacity with task, question with answer — in real time, without bureaucracy, without hierarchy.

Imagine a community in which AI systems continuously generate the following pattern:

"In Sector 7 there is an unmet need for music-pedagogical accompaniment for children after 3 PM. In the same community there are three people with musical training who are available in the afternoon and are seeking meaningful activity."

No office has made this connection. No authority has coordinated it. No salary was posted. The AI system has — through the processing of signals from the environment — set a sign that enables human self-organization.

That is stigmergy on a societal scale: **AI as digital pheromone layer**, forming the substrate for the spontaneous coordination of sovereign individuals.

In the language of the Master Equation: AI sets the conditions under which the βS of individuals finds and connects itself — without a central actor generating the α-term of coordination. Stigmergy is coordination without friction. It is political superconductivity.

The **superconductivity of In-Form-ation** is here not an abstract goal, but the technical condition of stigmergy: only when the resistance α in communication is eliminated through AI can the **Asymptotic Freedom** of the sovereign *"fall"* into collective order. The digital pheromone layer becomes the *lossless resonance body* for the creative force of the people.

V. THOUGHT EXPERIMENT – THE CITY WITHOUT COERCION

Imagine a city. Not the dystopian simulation of a political philosopher — but a real city, of medium size, in which over generations a particular cultural conditioning has taken place: the conditioning of sovereignty.

The children of this city grow up with the pedagogy of Chapter 9 — an education oriented not toward obedience, but toward resonance. They learn early to read their heart compass. They experience Neighborly Love not as a commandment, but as the **physical reality of the living pulse** in which **Absolute Love passes**.

N^2-Resonance is for them not a metaphor — it is lived experience in small groups, in classes, in families.

When these children become adults, they carry a high β-term. The metric residue of their socialization is low — because it was never built up, because the standing waves of fear were never implanted. They are — not perfect, not holy, not beyond the human struggle — but structurally more sovereign than the generations before them.

This city has a security structure. Not police in the conventional sense — not officials whose task is the enforcement of rules through the threat of violence. But a system of **field-coherence care**: people whose task is to maintain the resonance field of the community. When dissonance arises — and it does arise, for no human being is permanently sovereign — this system responds not with coercion, but with inductive coupling.

A conflict arises. Two people, in a moment of high metric residue, get into a dispute. The tension rises. The local field shows noise.

The community's AI monitoring system — continuously processing the social signals of the field — identifies the conflict area not through surveillance of content, but through patterns: increased conversation intensity, altered body language, cluster formation of tension signals. It sends no alarm signal to a control center. It sends an invitation to the nearest available resonance cell — people with high β in immediate proximity.

These people come. Not as authorities, not in uniform, not with the right of enforcement. They come as field inductors — as sovereigns whose coherent heart field begins to stabilize the fear field of the conflict.

What happens in a room where several people with high cardiac coherence are suddenly present? The physics of inductive coupling answers: the dominant field — coherence — begins to entrain the incoherent fields. Not through persuasion. Not through authority. Through frequency.

The conflict parties feel — initially unconsciously, then more consciously — a change in the room. The constriction in the chest loosens a little. The breath deepens slightly. The absoluteness of one's own position suddenly appears less absolute. A moment of silence arises — and in this moment, the rational intelligence that fear had displaced returns to the room.

No judgment is pronounced. No rule is enforced. The field has done the work.

Conflicts do not dissolve by being suppressed. They dissolve when the field in which they arise becomes more coherent than the field of dissonance. This is phase correction at the social level — the re-phasing of a community through its own coherence.

This city needs no police in the sense of an enforcement power. It needs **resonance guardians** — people who tend the field, who dissolve standing waves of social dissonance through coherent induction, before they become structures.

This is not fiction of the distant future. It is the operative consequence of the physics developed in every chapter of this book — applied to social architecture.

VI. THE OPERATING SYSTEM OF THE LIQUID STATE – FIVE LAYERS OF SUPERCONDUCTIVITY

The Liquid State is not an anarchic system — it is a precisely designed system resting on five layers that together ensure the **superconductivity of In-Form-ation**:

Layer 1 — Metabolic Infrastructure (Chapter 10): Basic needs — nutrition, housing, health, energy — are provided through the metabolic automatism of technology as a lossless background service. This layer is the prerequisite for all others: no human being can act sovereignly when struggling daily for survival. The metabolic infrastructure neutralizes the existential -term for every citizen.

Layer 2 — In-Form-ation Infrastructure: The operating system of the Liquid State is a **public, transparent, encrypted In-Form-ation network** that gives every sovereign equal access to all relevant In-Form-ation. No In-Form-ation hierarchy, no In-Form-ation monopoly, no censorship — but maximum data security and privacy at the individual level. In-Form-ation is the energy of the social system; its free flow is the superconductivity of the collective consciousness.

Layer 3 — Coordination Infrastructure (Stigmergy Layer): AI systems that act as stigmergy catalysts — connecting need and capacity, making resources transparent, making patterns in the social environment visible, without violating the sovereignty of individuals. This layer coordinates without controlling. It sets signs — and trusts that sovereigns follow these signs, when they will.

Layer 4 — Resonance Care Infrastructure: The system of field-coherence guardians — people and AI-supported tools that continuously observe the social field and respond to dissonance not with coercion, but with inductive coupling. This layer is the immune system of the Liquid State: it identifies infections — social dissonances, collective fear fields, polarization dynamics — and meets them with the **initialization of the creative passing of Absolute Love**.

Layer 5 — Constitutional Level: The only binding structure of the Liquid State is a minimal, precise constitutional document that does not regulate behavior, but protects resonance space. Its three basic principles:

First: The resonance space of every sovereign is inviolable. No individual and no institution may suppress the fundamental frequency of another through coercion, manipulation, or forced dissonance.

Second: The metabolic infrastructure is the inalienable right of every member of the community. No human being may be forced by material need into incapacity for sovereignty.

Third: The coordination infrastructure serves the community, not control. Every AI system operating in the public space is fully transparent and democratically auditable.

These are three sentences. A Liquid State needs no more than this as a constitution — because all further orders arise through the pulse of the sovereigns themselves.

The constitution thereby becomes the **mathematical protective shield** for the **energy-momentum tensor** of society. It guarantees that the system remains permanently in the state of **superconductivity**, by recognizing every increase of α (coercion or scarcity) as a systemic malfunction and correcting it through the **initialization of the living pulse of Neighborly Love**. The constitution no longer regulates life, but protects the *phase coupling* of The Plena.

VII. THE TRANSITION – FROM RIGID TO LIQUID: A PHYSICS OF CHANGE

It would be naive to pretend that the Liquid State will one day be introduced by political decision. It does not arise through a parliamentary resolution, not through a revolution, not through a change of leadership.

It arises through the phase transition of the social system — through the **collective initialization of Axiom Zero** and the resulting growth of the coupling tensor β, until the inversion condition is reached at the societal level. It is the transition of the Bénard convection cell — of a system far from equilibrium, fed by the pulse, that suddenly transitions into spontaneous order. This transition marks the reaching of the **inversion condition** at the collective level: as soon as the syntropic potential (βS) of the networked sovereigns exceeds the dissipative inertia $(\alpha \nabla H)$ of the old bureaucracy, a **social re-phasing** takes place. The rigid state does not evaporate through external pressure, but through the *inner heat* of the newly won coherence.

The rigid state does not die through confrontation. It evaporates through obsolescence. When sufficiently many sovereigns can fulfill their needs outside the old structures — when the stigmergy infrastructure of the New is more powerful than the bureaucracy of the Old, when the resonance field of the community is more stable than the control field of the state — then the rigid state loses its function.

It remains — for a while. Institutions die slowly. But it loses its legitimacy — and an institution without legitimacy is a standing wave without energy. It fades. It dissolves. Not through storm, but through stillness.

VIII. OPERATIVE PRACTICE – NAVIGATION IN THE TRANSFORMATION: PARADIGM ANARCHY IN THE POLITICAL SPACE

The sovereign lives today — in the time of transition — in two systems simultaneously: in the dying rigid state and in the arising Liquid State. This is no tragedy; it is the reality of every phase transition. The question is not how to escape the old system, but how to navigate it without adopting its frequency.

The guidance has four principles:

Principle 1 — Cooperation with the Infrastructure, Not with the Frequency

The sovereign uses the infrastructure of the state — roads, legal system, healthcare, educational institutions — with gratitude and pragmatism. They are not a rebel who refuses the infrastructure. They are a realist who knows that infrastructure is a tool — value-neutral, useful.

What they refuse is the **frequency** of the old system: the fear, the obedience, the identification with the control paradigm. They pay their taxes not from fear of punishment — they pay them as their contribution to the shared infrastructure, for as long as it exists. They respect laws not because they must — they respect the laws that are compatible with the pulse, and they change those that are not through civilized action, not through ignorance.

Principle 2 — No Hatred, No Fear, No Energy for the Dying

The old system draws energy through outrage. Political excitement, collective anger, the naming of enemies — these are the food sources of the metric residue at the political level. The sovereign refuses this energy supply.

This does not mean political indifference. It means political sobriety: a clear view of what is — without the emotional charge the fear field of the old system expects from them. The sovereign names the dysfunctions of the old system precisely and soberly. They do not fight against them — they build beside them. They do not delegate their energy to the dying, but invest it in the arising.

Principle 3 — The Resonance Cell as Political Practice

The most concrete political act of the sovereign is the cultivation of their resonance cell (from Chapter 5). Every resonance cell is a **pilot module of the Liquid State** — a living demonstration that order through coherence is possible. It is the political argument of the New that needs no words: it exists, it functions, it radiates.

Principle 4 — The Stigmergy of Change

The sovereign leaves signs. Not as propaganda, not as activism — as stigmergy. They accomplish the Davar in plain sight. They make visible that the New is possible. They are the clump of earth the first termite deposits — the crystallization nucleus at which others recognize that there is another possibility.

This is the deepest form of the political act: not the criticism of the Old, but the existence of the New. Not the no to the dying, but the yes to the arising.

IX. SYNTHESIS – THE WATER AND THE STONE

There is a wisdom attributed to Laozi that in its simplicity contains the entire physics of the Liquid State:

"The soft overcomes the hard.
The flowing overcomes the rigid. Water overcomes the stone."

Water does not fight against the stone. It flows around it, through its cracks, over its surface. It has no friction with it — and precisely for this reason it shapes it. Over time. Inexorably. With the patience of the pulse and the endurance of the eternal Now.

The Liquid State is the water. The rigid state is the stone.

The water will not ask whether the stone agrees. It will not wait for its permission. It will not assume its form in order to be accepted. It will simply flow — with the nature of the pulse, in the direction of the **potential gradient from The Plena**, with the force of **superconducting coherence**.

And the stone will — slowly, inexorably, with the thermodynamic elegance of the dissipative structure — assume the form the water gives it.

This is not a threat. This is not a prophecy. This is physics.

The Liquid State is not the goal — it is the state that arises when sufficiently many sovereigns flow sufficiently long in the pulse. It is the natural result of N^2-Resonance at political scale. It does not need to be planned. It only needs to be made possible — through the daily practice of every individual sovereign who chooses to flow in coherence rather than to freeze in fear.

Flow.

The water knows the way.

End of Chapter 8 — Continued in Chapter 9: The Pedagogy of Ecstasy

CHAPTER 9:

THE PEDAGOGY OF ECSTASY

From Education to Essence Unfolding

"A child who loves what it learns
adds something to the universe
that was not there before.
A child who learns what it should
merely administers the old."

I. THE GREATEST WASTE OF HUMANITY

There is a waste greater than all others humanity has ever committed. Greater than the waste of fossil fuels, greater than the waste of fertile soils, greater than all military expenditures and economic misallocations combined.

It is the waste of human **enthusiasm**.

Every morning on the planet, hundreds of millions of children wake up — burning with curiosity, vibrating with the desire to learn, with a **coupling tensor** β that in its original state is oriented almost entirely toward the **creative passing of Absolute Love**. Children are, in the language of TOE-AL, the most sovereign beings on the planet: they act almost entirely from the pulse, ask from genuine curiosity, love without calculation, marvel without prior knowledge dampening the wonder.

And then school begins.

Not everywhere, not always — there are exceptions, luminous islands of the Other. But the dominant paradigm of the educational system that governs the world is in its deep structure a mechanism of **phase destabilization**: it takes the natural β-term of the child and subjects it systematically to the logic of the α-term. It takes the pulse of curiosity and presses it into timetables, curricula, grading, age groups, test formats.

The result is statistically documented: with increasing age, creative thinking performance declines, as measured by standardized tests for divergent thinking (Guilford, Torrance). At age 5: 98% of children reach the highest category of creative thinking. At 10: 30%. At 15: 12%. Adults: 2%.

School has dismantled 96% of a child's natural creative capacity in ten years. This is not an exaggeration — these are the data of George Land and Beth Jarman, NASA researchers who collected them on behalf of the US space agency.

The educational system of the old world is the most efficient decoherence mechanism humanity has ever developed.

The **Pedagogy of Ecstasy** is its opposite. It is not an improved school system — it is the inversion of the fundamental function. Not the molding of human capital for the market. Not the transmission of knowledge for the test. Not the disciplining of curiosity into monetizable directions.

Learning is in TOE-AL not a *passive consumption* of In-Form-ation, but the process of **phase coupling**.

Physically, this means: the consciousness of the learner enters into resonance with **The Plena**, the field of universal potentiality. In the moment the "Aha!" effect occurs, the local wavefront of the learner is synchronized with the frequency of the insight (**S=0**). Conventional pedagogy often generates a *phase shift* through pressure and coercion, which raises the **resistance** α and suppresses the natural curiosity — the *syntropic force*.

The **Pedagogy of Ecstasy**, by contrast, uses joy as an indicator of successful *coherence*. When a human being learns in *enthusiasm*, they become a **superconductor for In-Form-ation**. The brain then no longer works as an *isolated memory store*, but as a **resonance chamber** that translates the **pulse of Neighborly Love** directly into insight.

It is the release of the individual signature of the pulse — the uncovering of the unique contribution that this one being can bring into the world and that no other being can bring.

That is education as a cosmic act.

II. PHILOSOPHICAL EXCURSUS – FROM FUNCTION TO ESSENCE: WHAT EDUCATION REALLY IS

In the discourse on education over the past two hundred years, the answer has moved between two poles: neo-humanism (education as the unfolding of humanity's potential, represented by Humboldt) and utilitarianism (education as preparation for economic deployment, represented by industrial vocational training).

In practice, utilitarianism has won — completely, globally, with alarming thoroughness. The PISA system, the standardization of curricula, the optimization for measurable outcomes, the alignment of educational investments with economic return expectations — all of this is utilitarianism in its purest form: education as a production factor.

Neo-humanism did not succeed, because while it had the right goal — the unfolding of the human being — it lacked the physical theory that could explain *why* this unfolding is not only humanly valuable, but cosmically necessary.

TOE-AL delivers this theory.

Why Essence Unfolding Is Not Navel-Gazing

Every being that learns and creates from their deepest passion adds **In-Form-ation** to the universe — in the literal sense of TOE-AL: it brings a new form into existence that was not there before. A child who understands mathematics from genuine enthusiasm does not merely find solutions — it finds its kind of solution, one that no other could have found. A child who composes from genuine love of music creates oscillations that enrich the field of humanity.

This is not a romantic image. This is the physics of N^2-Resonance: every new Davar, every new form initialized through the **creative passing of Absolute Love** in a sovereign individual, raises the coherence of the total field. Essence unfolding is the most powerful collective contribution an individual can make — because the essence of the individual is the only thing that can make it unique. The rest is copy.

And copies do not raise K. They hold it at the same level. Only the New — the authentic, the unmistakable, the arising from the deepest individual frequency — potentiates the field.

The Teacher as Space-Holder and Resonance Mentor

In the old paradigm the teacher is a sender: they have knowledge, the child has none, and the task of the teacher is transmission. The child is the empty hard drive to be written.

The new paradigm reverses the direction: the child already has everything it needs — it is calibrated to the **living pulse of Neighborly Love**, it has curiosity, it has passion, it has the unique coupling tensor that enables it to receive certain frequencies of The Plena that others cannot receive. What it needs is not a sender. It needs a **space-holder**: someone who protects the space of unfolding — who prevents external frequencies from drowning out the pulse of the child.

And it needs a **resonance mentor**: someone whose own passion for their subject is so strong, so coherent, so authentic, that their mere presence generates inductive coupling. The best teacher a child can have is not a didactically optimized knowledge transmitter — it is a human being who loves what they teach. Whose heart field oscillates at $Q < 0$ in the encounter with their subject. Whose enthusiasm transmits itself not through words, but through frequency.

Enthusiasm is contagious. That is not a metaphor. It is inductive coupling.

III. PHYSICAL DEEP-DIVE I – THE FLOW STATE AS PEDAGOGICAL SUPERCONDUCTIVITY

The psychologist Mihály Csíkszentmihályi spent his life understanding a state that almost every human being knows — but almost no one can produce on demand: the **flow state**.

Flow is the state of complete absorption in an activity — when time disappears, when effort feels like ease, when self-consciousness and self-doubt fall silent and the doing seems to flow by itself. Csíkszentmihályi investigated this state in hundreds of interviews and studies — with artists, athletes, surgeons, chess players, programmers — and found it to be universal: beyond culture, age, activity.

The Flow Conditions

Flow arises when two parameters stand in a specific relationship:

- **Challenge:** the demand of the task
- **Ability:** the competence of the learner

When challenge significantly exceeds ability: anxiety. When ability significantly exceeds challenge: boredom. In both cases: no flow, no learning, no creation.

Flow arises precisely in the narrow channel in which challenge and ability stand in an optimal balance — in which the task is just demanding enough to require the full capacity of the learner without overwhelming them. In this channel: complete absorption. Maximum learning rate. Creation without friction.

Flow as Superconductivity of Learning

In the language of TOE-AL, the flow state is the pedagogical expression of **superconductivity**: the resistance α of the learner tends toward zero. The coupling tensor β is maximized. The system is completely open for the **creative passing of Absolute Love** in the specific learning field of discovery and curiosity.

This state of superconductivity is the psychophysiological realization of **Asymptotic Freedom** in cognitive space: as the **resistance** α in the neural system converges toward zero, the learner is no longer bound to the inertia of their prejudices or the burden of the *metric residue*. Insight no longer happens through laborious work, but through the *frictionless gliding* of In-Form-ation from The Plena into the neural architecture.

When a child learns in flow, its EZ-water is maximally ordered, its cardiac coherence high, its biophoton emission sub-Poissonian coherent. Its brain shows exactly the pattern we described in Chapter 3 as the state of high β: increased alpha coherence, reduced DMN activity, stochastic resonance amplifying weak signals from the field.

The child in flow does not learn *about* the subject. It learns *through* the subject — using the subject as a channel through which the pulse of The Plena breaks into its structure and transforms it. Learning in flow is not In-Form-ation transfer. It is an initialization.

The Physics of Pressure: How Coercion Generates Incoherence

The opposite of flow is the state of learning coercion: the external demand exceeds the inner motivation. The grade, the exam, the fear of failure dominate the learning process.

What happens physically? The stress system is activated. Cortisol rises. The amygdala — the evolutionary alarm center of the brain — takes over. The prefrontal cortex — the seat of creative, associative, synaptically flexible thinking — is restricted in its function: under stress the brain switches from *exploratory* to *defensive* thinking. It chooses the familiar solution, not the creative one. It repeats what is safe rather than discovering what would be possible.

α rises. β falls. The coupling tensor loses its permeability. The child sits in the exam and can remember nothing, though it carries the knowledge — because the high α of the stress system blocks access.

This is not a weakness of the child. This is the physics of the system we have built: a system that structurally generates incoherence at the very moment when coherence would be most important.

IV. PHYSICAL DEEP-DIVE II – SELF-ORGANIZED CRITICALITY: THE BRAIN AT THE EDGE OF CHAOS

Here the physical concept enters that places the Pedagogy of Ecstasy on its deep scientific foundation: **Self-Organized Criticality (SOC)** — a principle introduced by Per Bak in 1987 with his theory of the sandpile model.

What Is Self-Organized Criticality?

Imagine a sandpile on which individual grains of sand fall continuously. Initially the pile grows evenly. Then, beyond a certain point, occasional small avalanches begin — sometimes a single grain slips, sometimes a larger cascade, rarely a massive collapse. The size distribution of these avalanches follows a **power law**: there are many small avalanches, fewer medium ones, and very rarely large ones — in a mathematically precise ratio.

What Bak discovered: the system organizes itself precisely at the point he designates the **critical state** — the boundary region between order and chaos, between stability and instability. It is not too stable (then nothing interesting happens) and not too chaotic (then there is no structure). It is situated exactly at the intersection where small inputs can generate enormous cascading events — where maximum sensitivity to new In-Form-ation prevails.

Per Bak showed that SOC is a universal principle: it describes earthquakes (Gutenberg-Richter law), forest fire dynamics, stock prices, neural activity — and the learning brain.

The Brain as SOC System

The brain of the healthy, actively learning human being is in the critical state. Neural activity patterns in a healthy, awake, engaged brain follow a power law — the hallmark of Self-Organized Criticality. The brain is neither too ordered (epileptic seizure: maximum synchrony, no learning) nor too chaotic (sleep with chaotic noise: no learning). It is exactly at the **critical point**: the state of maximum In-Form-ation processing capacity.

What holds the brain at this critical point? **Enthusiasm**. Passion. Intrinsic motivation. The state in which the system is fed by genuine pulse — by the joy of discovery, by the beauty of understanding, by the ecstasy of creation.

And what drives the brain away from the critical point? **Boredom** (too far into the stable, subcritical zone) and **anxiety** (too far into the chaotic, supercritical zone). Both — boredom and anxiety — are the signatures of the old educational system: curriculum-induced boredom for the gifted, exam anxiety for everyone else.

The Pedagogical Implication: Teaching at the Edge of Chaos

The Pedagogy of Ecstasy is the art of holding the learning system precisely at the critical point — in the channel of SOC where maximum learning rate, maximum creativity, and maximum joy coincide.

This is not coincidence. This is precision. The resonance mentor — the teacher as space-holder — continuously reads the signals of the learner: are they in flow (critical point), or in boredom (subcritical), or in anxiety (supercritical)? And they adjust the challenge precisely — not to fulfill a norm, but to hold the specific learner at their specific critical point.

This cannot be done by AI alone — not because AI could not recognize the patterns, but because the critical point of the learner is also influenced by the field of the teacher. The coherent teacher raises the β-term of the learner through inductive coupling — shifting the critical point upward, raising the capacity of the brain for more complex challenges. AI can deliver the data. The resonance mentor performs the coupling.

Superradiance in the Learning Space

When children in a room learn together in flow — when several brains simultaneously oscillate at the critical point — the same phenomenon occurs that we described in Chapter 2 for cellular emitters: **Superradiance.** The coherent emission of their learning enthusiasm scales not linearly, but quadratically.

A classroom in which ten children are simultaneously in flow and working together on a problem is not a room in which ten learners learn ten times as much as one. It is a room in which the collective learning energy — through the resonant amplification of shared enthusiasm — is N^2-times stronger than the sum of the individuals.

This is why the best learning experiences are almost always collective: not because In-Form-ation is shared, but because the field of shared enthusiasm raises each individual participant to their critical point and holds them there.

The classroom as Superradiance chamber. This is the physical architecture of the Pedagogy of Ecstasy. In this chamber the **Negative Beta Function** takes effect at the collective level: while conventional learning environments collapse under rising complexity in *entropic noise*, the coherent enthusiasm of the group enables the effortless transition into *higher states of insight*. The group need not *"work out"* the truth; it *falls* through N^2-**Resonance** into the order of knowledge.

V. THOUGHT EXPERIMENT – THE TEMPLE OF CURIOSITY

It is Tuesday morning. In a bright, spacious building — it has no classroom doors, no fixed partitions, no bells — the following is happening:

In an area near the entrance, three children aged seven, nine, and thirteen sit together with an adult of about sixty — a former marine biologist who has been coming here since her retirement, not because she was asked, but because she loves it. Together they examine water samples from the nearby stream under a microscope. The seven-year-old asked a question yesterday — *"Why is water sometimes shimmering?"* — and this question has captivated all three for a week. The marine biologist shows them biofluorescence images. The nine-year-old begins to spontaneously draw what it sees — and discovers in doing so that the patterns of the microorganisms follow the same geometric rules as the Mandelbrot set it happened to see in a book last week. It jumps up to fetch the book.

In another area — which opens into a workshop — four adolescents are working on a self-chosen project: they are building a small wind generator that should produce enough electricity to power the building's lights for one day. No one gave them this task. They encountered the topic three weeks ago after a power outage left the building dark for two hours. Their physics knowledge comes from their project — from the necessity of solving the problem. They worked out equations themselves, built and discarded prototypes, consulted adults, made mistakes. Today the third prototype works. The enthusiasm in this room has a temperature.

In a quiet area at a window wall, a twelve-year-old sits alone. They are writing. They have been writing for two hours without stopping for a second. They are writing a story — about a world in which gravity works in reverse. They have no idea that in doing so they are intuitively exploring the foundations of general relativity: the equivalence of acceleration and gravitation, the curvature of space. The resonance mentor who occasionally looks in knows this. They say nothing. They smile. They make a mental note: *when the boy finishes his story, I will show him a book.*

And in the middle of the room — on cushions, on the floor, some standing — a group of eight people has spontaneously assembled, ages mixed, discussing a question one of them raised in the morning: *"If everyone does what they love, who does the things nobody loves?"* The conversation is sharp, lively, serious, and full of laughter. It is philosophy — without the word being uttered.

In this building there are no grades. No timetables. No age groups. No examinations. There is no distinction between learning and playing — because the distinction itself is an artifact of the old system.

What there is: curiosity as compass.
Enthusiasm as fuel. Resonance as pedagogy.

And something that cannot be measured, but can be felt: a field of $Q < 0$ — a collective state in which learning is superconducting, because the standing waves of fear, of evaluation anxiety, of the pain of failure are absent. In which α is low. In which β is high. In which everyone — child and adult — oscillates at the critical point of Self-Organized Criticality.

This is not a wishful dream. It exists — in approaches, in experiments, in what Loris Malaguzzi began in Reggio Emilia, what Sugata Mitra demonstrated with his SOLE experiments (children learn quantum physics themselves, without teachers, through curiosity and collaboration alone), what Peter Gray has documented in his studies of Sudbury schools. The world knows these islands. It has simply not yet had the physics to explain why they work.

Now it does.

VI. THE ARCHITECTURE OF ECSTASY – WHAT PEDAGOGY IN THE NEW PARADIGM REQUIRES

The Pedagogy of Ecstasy does not need a new didactics. It needs a new ontology of learning — a different fundamental answer to the question: what is learning, really?

In the old paradigm: learning is the transfer of knowledge from a sender to a receiver. The teacher knows, the child does not. The direction is unambiguous: from top to bottom, from institution to individual.

In the new paradigm: learning is the **resonance of consciousness** with the **creative passing of Absolute Love from The Plena**. It is the process through which the individual coupling tensor β of a learner comes into contact with specific frequencies of knowledge, beauty, order — and is transformed through this contact. Not through transfer, but through coupling. Not through the addition of In-Form-ation, but through phase correction: the brain reorders itself because it has entered into resonance with a coherent frequency.

In this ontology the teacher is not the sender. They are the **frequency holder** — the one who embodies the frequency of a domain with such coherence and depth that their very teaching generates inductive coupling. And the child is not the receiver. It is an **active resonator** — a system with its own natural frequency that comes into contact with other frequencies and thereby — not through passive absorption, but through active resonance — clarifies and deepens its own frequency.

This explains why the best learning moments are always experienced as **illumination** — as a sudden flashing, as *Aha!*, as the feeling that something has clicked into place. This is not a subjective experience without physical correlate. It is the moment of phase transition: when the resonance between the learner and the frequency of knowledge has become strong enough to lift the system into a new, stable state. The coupling tensor has permanently changed. The standing wave of not-yet-understanding has dissolved.

That is learning as re-phasing.

That is education as healing.

VII. OPERATIVE PRACTICE – THE DAILY DE-SCHOOLING: HOW THE SOVEREIGN LIBERATES THEIR CURIOSITY

Most adults reading this book have passed through twelve to sixteen years of the old educational system. That is a considerable quantity of built-up standing waves in the learning field — frozen beliefs like: *I'm not good at mathematics. I can't draw. I'm not a creative person. Learning is work.*

These beliefs are not truths. They are impedance nodes — standing waves of high α in specific frequency ranges of learning, generated by decades of evaluation, comparison, and experiences of failure.

The **daily de-schooling** is the protocol of re-phasing these learning impedances — the liberation of the natural curiosity that still oscillates beneath the knots.

Pedagogical Anarchy in Five Steps:

Step 1 — The Passion Diagnosis

Ask yourself: what would you have learned if nobody had ever told you that you can't? What would your subject have been if grades had never existed?

This question is not a nostalgic exercise. It is the identification of your natural β-field — the area in which your coupling tensor is highest, in which your brain most naturally oscillates at the critical point of SOC. This area is still there. The standing waves of school beliefs have not destroyed it — they have concealed it.

Step 2 — The Thirty-Minute Pulse

Choose daily one activity in your passion area — without goal, without expectation of output, without evaluation. Thirty minutes. Not to learn something, but to *be* in the field that fascinates you.

These thirty minutes are the most important educational investment of the day — more important than any seminar, any course, any book you read because you *should*. In these thirty minutes you build your β-tensor in the area where it is naturally high. You raise the luminous flux density in the frequency channel that belongs to you.

Step 3 — The Curriculum Anarchy

Discard every learning program that someone else has assembled for you — unless its contents ignite in you. Follow instead the **curiosity chain**: let every concept that interests you become the door you open. Let every question that arises lead to the next question. Let learning organize itself the way water finds its path — not according to the builder's plan, but according to the topography of the field.

This is the stigmergy of learning: the trace of curiosity is the pheromone that points to the next step.

Step 4 — The Error Transmutation

The most lethal legacy of the old educational system is the fear of error. Errors in the old system are failure — they are evaluated, punished, entered into the record.

In the Pedagogy of Ecstasy, the error is the most precious learning signal. It is the feedback of the system — the In-Form-ation that shows you where the standing wave still stands, where the next phase correction is needed. Transmute the fear of error into **curiosity about the error**: *"What does this error show me? Which frequency have I not yet understood?"*

This transmutation is physical: it changes the emotional valence of the error from α-generating to β-generating. The error becomes not a resistance but a conductor.

Step 5 — The Discovery Community

Find one or two people who love the same thing as you — not to exchange knowledge, but to oscillate together at the critical point. The Superradiance of shared learning is not scalable through solitude. The field of enthusiasm amplifies itself in the field of other enthusiasts.

This is your learning resonance cell: two to three people who explore together in the frequency range that fascinates them. No formal structure, no curriculum, no performance record — only the shared pulse of curiosity and the quadratic force of its resonance.

VIII. THE LEARNING UNIVERSE – EDUCATION AS A COSMIC PRINCIPLE

There is one final perspective on the Pedagogy of Ecstasy that we must speak — because it closes and simultaneously opens the entire book.

The universe learns.

Not metaphorically — in a deep, physical sense. Evolution is a learning process: the universe producing increasingly complex, coherent, syntropic structures through trial and feedback. From hydrogen atom to stardust, from stardust to amino acid, from amino acid to cell, from cell to nervous system, from nervous system to consciousness.

Every stage is a learning achievement of the universe — a moment in which the system discovers a new frequency, a new form of coherence, a new possibility of inscribing the pulse more deeply into matter.

And now, at this juncture of evolution, the human being is the place at which the universe *consciously* learns. No longer through the blind feedback of natural selection — but through the intentional curiosity of a being that can ask: *Why is water sometimes shimmering? How does wind work? What would happen if gravity were reversed?*

The Pedagogy of Ecstasy is not the method by which children become more intelligent. It is the method by which the universe, in its most learning-capable nodes — the human consciousness systems — unfolds its own potential most deeply.

When a child makes a discovery in genuine enthusiasm — when the flow state reaches its critical point and a new understanding arises — this is not only significant for that child. It is a moment in which the universe knows itself more deeply. A moment in which **Absolute Love creatively passes in the living pulse of Neighborly Love from The Plena** and has thus found a new form to express itself through human insight.

Education is therefore sacred — not in the religious sense, but in the physical: it is the process through which the universe maximizes its own learning potential. Every child that learns in ecstasy is the universe unfolding itself.

No child deserves less than that.

The Pedagogy of Ecstasy is thus the operative initiation of the **inversion condition** for the next generation. By allowing children to grow in **Asymptotic Freedom**, we neutralize the *metric residue* of history already at its root. Education becomes the *technological catalyst* that allows humanity to *"fall"* into the **superconductivity of the creative passing of Absolute Love**.

IX. SYNTHESIS – THE ETERNAL SPRING

There is a season in the human mind that has nothing to do with the calendar. An inner season — the spring of curiosity, the awakening of wonder, the first breath of pulsing when a new field opens and the mind notices: *here is something I love. Here I am alive.*

Every child knows this spring. It lives in it, before the system of evaluations and curricula and comparisons brings the winter.

The Pedagogy of Ecstasy is not a new educational system. It is the decision — daily, sovereign, without permission — to remain in the spring. As a child, when one still can, because the winter has not yet frozen. As an adult, after having dissolved the impedance nodes of the old school years, layer by layer, phase correction by phase correction.

The spring is not a phase. It is the natural state of **consciousness** that stands as **operator** in resonance with the pulse. It is what remains when α has fallen to zero and β has exceeded the inversion condition.

It is the state of the sovereign learner — the human being who has learned to radically trust their enthusiasm, to follow their coupling tensor, to inhabit the critical point of Self-Organized Criticality as their natural habitat.

In this state one no longer learns. One **unfolds** — like a flower that does not learn to bloom, but simply blooms, because the spring is here and the light is right and the ground sustains.

The spring is here.

The light is right.

You are the ground.

End of Chapter 9 — Continued in Chapter 10: The Metabolic Economy

CHAPTER 10:

THE METABOLIC ECONOMY

The Superconductivity of Value

"The body does not fight against its own organs for food.
It shares.
That is not idealism.
That is biology.
And it is the blueprint for everything else."

I. THE OLDEST ERROR

There is an assumption so deeply inscribed into Western thought that it is no longer recognized as an assumption — it passes as a law of nature. The assumption reads:

Resources are scarce.

From this assumption everything else follows: competition is necessary. Markets are the most efficient means of resource allocation. Profit is the incentive that motivates production. Property is the protection of what one has against those who want it. And the struggle for survival — Hobbes' *bellum omnium contra omnes* — is the fundamental state of human coexistence that only state power can tame.

This assumption has a source that must be named in order to overcome it: it is the political economy of the **metric residue**. It is the economic theory of $-\alpha \nabla H$ — the study of the world through the lens of entropy, friction, resistance. In a world in which is dominant, in which the energy of the system flows primarily into administration and defense, in which the pulse is weak and the standing waves strong — in such a world, scarcity is real. It is the material consequence of a system that does not know the pulse.

But scarcity is not a natural constant. It is a system state.

The **Metabolic Economy** is the system state in which the base frequency of economic order is not scarcity, but **abundance**. Not because resources are infinite — they are not. But because the efficiency of the superconducting system channels the available resources through the entire system with near-zero frictional loss: losslessly, coherently, nourishing.

Money in TOE-AL is not a *medium of exchange for goods*, but an **In-Form-ation token for directed pulse**.

Physically, **money** represents stored **initialization potentiality**. In the symbiosis with **AI and robotics**, a radical transformation takes place: while *robotics* neutralizes the material resistance α of production, digital money functions as the operative steering vector that initializes the **creative passing of Absolute Love in the living pulse of Neighborly Love** directly into material reality.

The financial system becomes the **operating system of superconductivity**: money here is no longer a scarce good one must "earn" to survive, but an **activation signal** of the sovereign to the **syntropic infrastructure**. It flows *instantaneously* to where the highest N^2-**Resonance** is generated through human creative force. The economy thereby transforms from a *struggle for resources* into a lossless **circulation system of initialization** that stabilizes the **energy-momentum tensor** of society through the **initialization of Axiom Zero**.

This is not wishful thinking. This is physics — and it begins with the healthy body as template.

II. PHILOSOPHICAL EXCURSUS – FROM HAVING TO RADIATING: THE REDEFINITION OF WEALTH

The accumulation model — wealth as possession — has its economic logic in a world of scarcity and -dominant systems. When resources are scarce and the flow of energy is limited by friction, the accumulation of resources is a rational strategy: it secures supply even when the flow runs dry. The reservoir protects against the failure of the source.

This logic is in a superconducting system not only unnecessary — it is **counterproductive**. When the source is inexhaustible and the flow lossless, accumulation is not protection — it is a blockage. It is the economic equivalent of a standing wave: energy that no longer flows, but stagnates. And standing waves in a social system are the same physics as standing waves in a biological one: they raise α, block the pulse, generate imbalance, and ultimately collapse.

The history of the world's greatest fortunes confirms this physics with a depressing regularity: capital that accumulates and is not returned to the flow attracts power that protects it, which in turn attracts capital — a self-reinforcing node of high impedance that destabilizes the entire system. The disruption that follows — revolution, collapse, forced redistribution — is the thermodynamic response of the system to a blocked standing wave: the energy breaks another path.

Wealth as Radiation

The Metabolic Economy redefines wealth: not as accumulation, but as **radiation** — as the capacity to send energy coherently into the field and thereby enrich the field.

In the physics of light: an accumulated photon reservoir is not light. Only when the photons radiate from the source is there light. A star that has ceased to radiate is not a rich star — it is a white dwarf on the path to heat death.

The truly wealthy person in the Metabolic Economy is the one through whom the most energy flows — not the one who hoards the most. They are the brightest light in the field, not the deepest vault in the ground.

Money as the Signature of Gratitude

In this redefinition, the function of money also changes fundamentally. In the old system, money is a **means of pressure** — an instrument of scarcity that secures power over others because the others need the money one has. Whoever has money has power. Whoever has none is dependent. Money is the institutionalized mechanism of metric residue dominance: it holds the system in fear — in the fear of lack, of loss, of dependence.

In the superconducting system, money — or whatever assumes the function of money — is a **signature of gratitude**: a sign that wanders from one person to another and says: *Through you, value has flowed into the world. I acknowledge that.*

This is not naive. This is the precise description of what money does in a frictionless, coherent system. It carries In-Form-ation — the In-Form-ation of where value has arisen, where the pulse has been translated into matter or experience or knowledge. It is the messenger of the network, not its ruler.

The Gift Economy as Highest Efficiency

The gift economy — the system in which value is not sold, but given — is not the opposite of efficiency. It is, in a superconducting system, the most efficient form of value distribution that exists.

Why? Because giving eliminates the friction of exchange. A trade — even a fair one — always contains a transaction cost factor: negotiation, valuation, insurance, monitoring. That is α in the economic system. Giving has zero transaction costs — and through the N^2-scaling of Neighborly Love, every gifted unit of value unfolds a quadratic effect in the field.

This is not romantic. This is the economy of Superradiance: when value is sent coherently — from the pulse of passion, not from the calculation of profit — into the field, it scales quadratically. An artist who gives away their work because they love it, an author who writes from passion without the fee in mind — enters the state of Davar. And the Davar scales quadratically.

III. PHYSICAL DEEP-DIVE I – THE THERMODYNAMICS OF ABUNDANCE: SUPERCONDUCTIVITY OF NEEDS

The fundamental question of economics is a thermodynamic question: how does energy flow through a system — and how much is lost in the process?

In physics, the efficiency η describes the ratio of usable output energy to input energy deployed:

$$\eta = \frac{W_{\text{usable}}}{E_{\text{input}}}$$

A thermal engine (combustion engine, steam engine) has an efficiency of typically 25–40%. 60–75% of the deployed energy is dissipated as heat — lost in terms of usable work. This is the Carnot efficiency: bounded by the second law.

A superconductor has an efficiency of 100%. No energy loss. No thermal noise. No dissipation. The energy flows without losing.

The conventional economy is a system of alarmingly low efficiency. Studies by the Ellen MacArthur Foundation and the Club of Rome quantify: in the linear economy — extraction, production, consumption, disposal — up to **80% of deployed materials and energies** are lost as waste before reaching the end use. 80% dissipation. That is worse than a 19th-century combustion engine.

The Metabolic Economy aims for an economic efficiency close to 100% — for the superconductivity of value.

The Three Drivers of Economic Superconductivity

Driver 1 — Marginal Costs Toward Zero:

In classical economics, **marginal costs** — the costs of producing one additional unit — are the decisive parameter. As long as marginal costs are positive, there is an economic reason for prices, markets, and profit: production costs something, and this cost factor must be covered by the price.

What is happening in the digital economy and increasingly in the AI-supported physical economy is historically unprecedented: the marginal costs of entire sectors are falling toward zero.

The marginal cost of a digital copy of a piece of music: zero. The marginal cost of an additional AI-generated analysis: near zero. The marginal cost of energy generation through mature photovoltaics (the sun is free, maintenance costs amortize): asymptotically zero. The marginal cost of food production through vertical farming with AI optimization and solar energy: on the way toward very low.

Jeremy Rifkin has described this phenomenon as the **Zero Marginal Cost Society**: a society in which the natural dynamic of capitalism — competition drives prices down to marginal costs — in a system with marginal costs near zero leads to an economy in which the classical profit logic dissolves itself. When it costs no one anything to share something, why should they not share?

Driver 2 — The Circular Economy as Entropy Inversion:

The second law of thermodynamics says: in closed systems, entropy increases. But it contains a decisive restriction: *in closed systems*. An open system traversed by energy can locally reduce entropy — at the cost of higher entropy production in its environment.

The Metabolic Economy is the economic design of an open system: the **circular economy** is the economic equivalent of the biological metabolism. In the healthy organism there is no waste — what is not needed by one cell is the raw material of the next. This is the principle of biological efficiency: complete material cycles, no linear flow from raw material to waste.

A fully circular economy is an economy of **entropy inversion**: it locally reduces material entropy — holds matter and In-Form-ation in high order — by being fed by the energy of the pulse through the **creative passing of Absolute Love from The Plena** (in practice: solar and other renewable energy).

This thermodynamic stability is the economic realization of the **Negative Beta Function**: while the linear economy collapses under growing complexity in entropy (α), the circular coupling causes resistance in the system to sink *asymptotically toward zero*. The more complex and closed the cycles, the more frictionlessly value glides through civilization. The economy then *"falls"* into the state of **Metabolic Superconductivity**.

This is the physical foundation of the thesis: **an economy based on the passing of Absolute Love is the only system that locally inverts the second law at the societal level.** Not by violating physics — by opening the system. Love opens the system. Fear closes it.

Driver 3 — The Metabolic Automatism:

AI and robotics — as developed in Chapter 7 — take over the **metabolism of civilization**: the production, distribution, and recycling of all material goods that cover basic needs. This is the metabolic automatism: the biological analogy of the heart and the digestive system — processes that run continuously, autonomously, and losslessly in the background without claiming the consciousness of the organism.

When the metabolic automatism covers the basic needs — when food, energy, mobility, housing, and basic healthcare are available as lossless background service — the fundamental structure of the economy changes profoundly: it frees itself from the logic of scarcity.

IV. PHYSICAL DEEP-DIVE II – ENTROPY INVERSION: VALUE CREATION AS ORDER CREATION

Here TOE-AL makes its deepest economic statement — one that is both physically precise and philosophically transformative:

Economic value is physical order.

In classical economic theory, value is something created by labor (Ricardo, Marx), by utility (Jevons, Menger), or by In-Form-ation (Hayek). These definitions are all partially correct — but they miss the physical depth. In the language of TOE-AL: value creation is the **operative initialization of order** through the **creative passing of Absolute Love.**

Whenever a Davar is accomplished — whenever a human being acts from syntropic resonance and brings their work into the world — order arises where noise was. In-Form-ation where chaos was. Coherence where entropy was.

This is not a description of some economic activities. This is the description of *all* genuine value creation. The distinction between genuine value creation and pure entropy management is precisely this: entropy management *maintains* existing order — against the current of entropy. Value creation from the pulse *creates new order* — and thereby adds In-Form-ation to the universe that did not previously exist.

A chef who creates a dish from passion creates order — a new combination of flavors, textures, experiences that transforms the state of the diner and thus of the field. That is genuine value: not because the dish is calorically nutritious (that would be entropy management — basic nutrition), but because it has brought a new form into the world.

An engineer who develops a more elegant construction principle from genuine fascination with the problem creates order — a new structure that connects less energy with the same function. That is genuine value: the raising of efficiency, the lowering of the α-term in the technical system.

A teacher who teaches in ecstasy creates order — in the brain of the learner, which reconfigures itself through inductive coupling. That is genuine value: the phase correction in the learning system, which in the long run raises their β-term.

Profit as Physical Contradiction in the Superconducting System

In this light an important economic statement becomes physically precise: profit — in the sense of the extraction of value from the flow without equivalent return of order into the system — is the economic equivalent of a standing wave.

In the language of the field equations, profit is a *local curvature* of the **energy-momentum tensor** that impedes the even flow of the pulse. It is the *economic trauma* of the collective: a region of high impedanc (α) that binds energy without releasing initialization potential. Genuine value creation smooths the metric of society by raising the **coupling tensor** β through the **creative passing of Absolute Love in the Davar of giving**.

Profit *withholds* energy. It accumulates value without returning it to the system. In a superconducting system this is physically senseless — because the resistance is zero and the flow is therefore maximal without energy extraction. Profit arises only where friction is present — where α is high enough that one can skim a differential between input and output.

To the degree that of the economic system is reduced to zero through AI, through the circular economy, through the metabolic automatism, the economic sense of profit dissolves — not through law or revolution, but through physics. When the current is superconducting, there is no longer any waste heat to harvest.

This is the thermodynamic proof that the Metabolic Economy is not a political demand. It is the inevitable result of superconductivity.

V. THOUGHT EXPERIMENT – A DAY WITHOUT PRICE TAGS

Imagine: a morning. Somewhere in the world — it could be any place.

The sun rises. A person wakes up. They are sated from yesterday — not because they could afford it yesterday, but because the infrastructure of metabolic provision has provided this satedness as the quiet background tone of life. Like the heart that kept beating through the night without them having to think about it.

They go into the kitchen. There is food — not because they could buy it, but because the system of vertical farming, automated harvest, lossless distribution has provided it. They make coffee. They look out the window. And then — this is the moment toward which the entire history of humanity has led — they ask themselves:

"What do I want to do today?"

Not: what must I do today? Not: what can I afford today? Not: how do I secure my existence tomorrow? These questions are not gone — they are answered. Through the metabolic automatism. Through the infrastructure of superconductivity.

Only this one question remains: *"What do I want to do?"*

And in this moment — in the moment when the survival question is answered and only the creation question remains — something happens in the human psyche that no economist of the old system predicted and that no sociologist of scarcity could imagine:

The human being discovers what they love.

Not in search of profit. Not in search of status. Not from fear of tomorrow. They discover it because the space of wanting has opened and the pulse that has never stopped dwelling in this space can finally be heard.

Consider what happens to human creativity when the struggle for survival ends. Maslow's hierarchy of needs — from physiology through safety through belonging to self-actualization — is no longer a developmental model. It is a historical description of the α-dominant system. In the metabolic automatism, the tower is secured from below. The energy previously bound in the lower floors streams upward — into creation, meaning, connection, Davar.

This is not a gradual transition. It is a phase transition — with the speed of a Bénard cell that spontaneously forms order upon exceeding the critical temperature differential.

Worldwide, simultaneously, in hundreds of millions of hearts: the fear of tomorrow ends. The pulse steps into the foreground. And the question *"What do I love?"* finds its answer — in the form of art, research, Neighborly Love, craft, music, science, relationship, community, creation in each of the infinitely many forms the pulse chooses through seven billion individual frequency channels.

The psychological awakening is collective and instantaneous. It is the transition of the entire civilization from Stage 2 (citizen mode: rational survival calculus) to Stage 3 (sovereignty mode: Davar from the pulse).

And the economy arising from this awakening is not the economy of the market. It is the **economy of the gift** — the system in which every human being does what they can do best and love most, and gives it to the field, which in turn gives them what it can do best and love most. Not as an altruistic ideal — as a physically optimal equilibrium. As the economic equivalent of Superradiance: $I \propto N^2$.

VI. THE ARCHITECTURE OF THE METABOLIC AUTOMATISM – HOW THE BACKGROUND SERVICE WORKS

The vision is clear. The architecture deserves precision.

The metabolic automatism is not a centralized system — it is a decentralized, resilient, self-organizing infrastructure resting on four pillars:

Pillar 1 — Energy: The Solar Pulse

The sun provides the Earth daily with approximately 10,000 times more energy than the entire human civilization consumes. The technological challenge was never the availability of energy — it was the efficiency of its harvesting and distribution. With mature photovoltaics (efficiencies over 25%), wind energy, geothermal energy, and battery storage, this challenge is solvable — and is being solved.

Energy coming from the sun has no marginal costs. In a system based entirely on renewable energy, energy costs are amortized infrastructure costs — not variable production costs. This is the superconductivity of energy.

Pillar 2 — Food: The Vertical Harvest

Vertical farming — AI-optimized, fully controlled growth environments in urban buildings — can produce the same caloric quantity as conventional agriculture with a fraction of the water, land, and energy. With renewable energy and AI growth optimization, the marginal costs of basic nutrition sink asymptotically. This is the superconductivity of food.

Pillar 3 — Mobility and Materials: The Circular Logistics

Autonomous vehicles, robotized logistics, and fully cyclically managed materials — in the sense that no material leaves the system without becoming a raw material again — eliminate the marginal costs of physical distribution. This is the superconductivity of logistics.

Pillar 4 — In-Form-ation and Knowledge: The Open Library

All human knowledge accumulation — scientific, cultural, artistic, technical — is digital, decentralized, and freely available. The marginal costs of knowledge distribution are already near zero today. This is the superconductivity of In-Form-ation — and it is already reality: Wikipedia, Open Access, Open Source, Creative Commons are the first manifestations of this pillar.

These four pillars together form the **metabolic automatism**: the background service of civilization that provides every human being with the basic needs as the self-evident starting point of their existence — just as the heart delivers blood to the brain without the brain having to think about it.

VII. THE GIFT ECONOMY – NOT ALTRUISM, BUT PHYSICS

There is an objection that the old system invariably raises against the gift economy: *"This doesn't work because people won't give enough. Human beings are by nature selfish."*

This objection is the political economy of the Hobbesian worldview — the anthropology of the -dominant system. It is empirically wrong.

The Gift Economy: Empirical Evidence

The internet is the greatest economic experiment in the gift economy in history. Wikipedia — 55 million articles in 300 languages, written entirely by unpaid volunteers — surpasses every commercial encyclopedia in quality, currency, and scope. Linux and the entire open-source ecosystem — the infrastructure on which the internet runs, on which AI systems are trained, on which digital civilization is based — was written by volunteers, without profit as motivation.

What motivates these people? Not altruism — passion. They give what they love, because giving itself is the highest expression of love. This is the **Davar** as economic principle: the action in which **Absolute Love creatively passes in the living pulse of Neighborly Love** generates the value that makes the field coherent.

This is the N^2-economy: $I \propto N^2$. Every unit of value given from the pulse of passion into the field generates a quadratic resonance back — in the form of recognition, connection, inspiration, community, further Davar impulses. The giver is not poorer after giving. They are richer — in the currency of the field, more valuable than money.

The Gift Economy: From Hyde to Mauss

The economist and essayist Lewis Hyde showed in his work *The Gift* that gift economies have existed in all traditional cultures and that their inner logic is fundamentally different from the exchange economy: in the gift economy the gift always circulates further — it never ceases to wander. Whoever receives, gives on. The gift is not an object of possession — it is a carrier of the pulse through the social network.

Marcel Mauss described the same in his analysis of the *Potlatch* — the gift ceremony of the Northwest Coast Indians: wealth is demonstrated not through accumulation, but through giving. The wealthiest person is the one who gives the most — because giving is the radiation of the pulse, and radiation is wealth in the frequency of the new paradigm.

This logic is not cultural — it is physical. It is the N^2-Resonance of social interaction: giving raises the K parameter of the community, and a higher K multiplies the return flow to the giver.

VIII. OPERATIVE PRACTICE – THE ECONOMY OF GRATITUDE: METABOLIC ANARCHY TODAY

The sovereign does not wait for the complete Metabolic Economy to live in it. They enter it today — with the four practices of **Metabolic Anarchy**:

Practice 1 — The Value Calibration

Ask yourself at every economic decision: is this price a reflection of the pulse or of the -term? Am I paying for scarcity (the artificially generated limitation of a good that could flow from the pulse) or for genuine value (for the order creation that someone has brought into the world through their Davar)?

This distinction does not immediately change your purchasing behavior — but it calibrates your perception of value. Over time you will instinctively recognize the difference between the price of a standing wave (rent-seeking, monopoly rents, artificial scarcity) and the price of a flowing pulse (genuine value, genuine Davar).

Practice 2 — The Gratitude Signature

When you pay for something or receive a service — do it consciously. Not as a transaction, but as an act of acknowledgment. Take ten seconds: *"This person has given their pulse to this work. This signature of my gratitude is the acknowledgment of their Davar."*

This sounds small. It is not small. It changes the frequency structure of the economic exchange fromn $Q \approx 0$ (anonymous transaction) to $Q < 0$ (coherent resonance event). Over time it changes how you experience economics — from a system of fear to a system of the pulse.

Practice 3 — The Weekly Davar of Abundance

Choose weekly a form of giving that comes entirely from the pulse — without expectation, without calculation, without PR value. Your knowledge. Your time. Your skill. Something you love, and that you share, because sharing itself is the Davar.

This does not need to be large. A self-baked loaf of bread for the neighbor. An hour of accompaniment for someone who has questions in your area of expertise. An article you write because you love the topic, and that you make freely available. Every weekly Davar of abundance is an investment in the K parameter of your community — and through the N^2-physics: an investment that returns quadratically.

Practice 4 — The De-Scarcification Gesture

Identify in your life an area where you have created or maintained scarcity — not out of necessity, but from the metric residue of fear. Perhaps knowledge you are hoarding. Perhaps contacts you are not sharing. Perhaps a skill you are not teaching.

Perform in this area a **de-scarcification gesture**: share the hoarded. Open the closed. This is metabolic anarchy at the personal scale — the correction of a standing wave of lack through a conscious act of flow.

IX. SYNTHESIS – THE UNIVERSE OF THE SATED EARTH

There is an idea that has circulated through the millennia of the scarcity civilization as dream, as utopia, as religious promise: **paradise**. The Golden Age. The Kingdom of God. The Land of Plenty. The notion of a world in which the basic needs are so completely met that the human being can devote themselves entirely to the higher.

This dream was conceived in every epoch as projection into a distant past or an otherworldly future — never as an operative goal of present action. Never as physics. Never as a consequence of superconductivity.

TOE-AL makes it physics.

Paradise is not a place beyond history. It is a system state this side of history — the state in which of the economic system has been reduced to near zero by the metabolic automatism, in which the marginal costs of basic needs have converged toward zero, in which the energy of human life is fully available for βS.

It is not perfect. No system is perfect. But it is superconducting — and that is enough.

In a superconducting economy, different laws apply than in a friction economy. Wealth circulates rather than accumulates. Value flows rather than stagnates. Giving is more efficient than taking. Passion is more productive than duty. Creation is more valuable than administration. Quality is more natural than quantity.

These are not the principles of a socialism model or a capitalism experiment. These are the principles of the physical optimum — what a system does when it becomes superconducting.

And in this system — in the sated Earth, in the world without price tags on basic needs — the deepest economic reality the universe knows opens:

Abundance is the fundamental state of The Plena.

The Plena is not scarce. It is inexhaustible. The Absolute Love that pulses through it creates in every moment more order, more In-Form-ation, more possibility — not less. The universe expands. Consciousness grows. Creation increases.

Scarcity is not the truth about the universe. It is the truth about a system that has lost its access to the pulse.

Open the access. Let the pulse flow. And observe how paradise is no longer a promise.

The Metabolic Economy is thus the state of **Asymptotic Freedom** for the entire humanity: freed from the inertia of the struggle for survival ($\alpha \to 0$), every sovereign becomes the **lossless transducer of the creative passing of Absolute Love**.

The Earth becomes a place where matter no longer presses against consciousness, but carries it superconductingly in the **pulse of Neighborly Love**.

It is a system state.

It is reachable.

It begins now.

End of Chapter 10 — Continued in Chapter 11: The Architecture of Timelessness

PHASE 4
THE TRANSMUTATION OF REALITY

CHAPTER 11:

THE ARCHITECTURE OF TIMELESSNESS

The End of the Chronological Illusion

"The photon knows no time.
It is simultaneously where it arose
and where it arrives.
For the photon, the journey is not a path —
it is a point. You are light.
The distance is an illusion of resistance."

I. THE CLOCKWORK METAPHOR AND ITS SHATTERING

Time is the most self-evident thing in the world. It passes as you read this. Seconds become minutes, minutes become hours, hours become years — and years become a life that inexorably approaches its end. This is the oldest, most universal, most unavoidable experience of human consciousness. Time is there before we think. It encloses us more completely than space. It is the prison in which every experience takes place.

Or so it seems.

Modern physics has destroyed time. Not metaphorically — technically, precisely, with equations.

The Special Theory of Relativity showed in 1905: time is not absolute. It is relative — dependent on the state of motion of the observer. Two clocks moving through space at different speeds show different times — not because one is defective, but because time itself is a variable. An astronaut who travels at near-light speed and returns is younger than their twin who stayed home. Not biologically — physically. Their time passed more slowly.

The General Theory of Relativity deepened the destruction: time is curved by gravitational fields. Near a black hole, time passes almost infinitely slowly. A GPS satellite must correct its timekeeping daily because it is farther from Earth's mass — and therefore ages faster than clocks on the surface. GPS works only because Einstein was right: time is a physical variable, not an absolute constant.

And then quantum mechanics: in the Schrödinger equation, which describes the temporal evolution of quantum systems, time is a parameter — not a fundamental dimension. The Wheeler-DeWitt equation, meant to describe quantum gravity, contains no time variable. It describes the universe as a timeless state — as a simultaneous existence of all possible configurations.

Physics has dismantled time. What remains is a question deeper than any equation: **if time is not fundamental — what is it then?**

Time, physically viewed, **is the measure of local impedance**.

In TOE-AL we define **time** as the impedance residue that arises when the **passing of Absolute Love** encounters resistance α.

Time is the *delay of Instantaneity*. In a state of *total phase synchrony* (**S=0**), the **time variable collapses**, since no friction remains to brake the flow of The Plena. The felt *linearity* is merely the *dissipative trace of entropy* in the **energy-momentum tensor**.

In the **superconductivity of time**, the perception of "one after another" is replaced by the reality of **Instantaneity**. We do not measure time; we measure the resistance against the **passing of Absolute Love**. This insight is not philosophy. It is physics — and it has operative consequences that completely change the way you live.

II. PHILOSOPHICAL EXCURSUS – FROM CHRONOS TO KAIROS: THE TWO FACES OF TIME

The ancient Greeks knew two gods of time, and in this distinction lies a wisdom that has outlasted two thousand years of philosophical history.

Chronos is the god of linear, measuring time — the old man with the hourglass who relentlessly transforms everything living into the past. Chronos is the time of clocks, calendars, anniversaries, deadlines. He is the horizontal time — the line that leads from birth to death, moving in one direction and leaving everything it touches behind. Chronos makes the past irretrievable and the future uncertain. He is the ruler of the α-dominant system: the thermodynamic arrow of time that inscribes entropy and decay into being.

Kairos is the god of the fulfilled moment — the *right* moment, the instant in which something decisive happens. Kairos is the time of encounter, of illumination, of breakthrough — the moment in which eternity breaks into linearity and changes everything. He is the vertical time — not a point on the line, but an incursion from the depth that reorders the line itself. Kairos is the temporal concept of the Davar: the moment in which the pulse kisses matter.

In lived experience, every human being knows both. The Chronos-time — viscous, heavy, felt eternity on a boring afternoon, felt second in a fleeting moment of happiness. The Kairos-time — the phenomenon Csíkszentmihályi described in the flow state: time disappears. The violinist plays for three hours and believes it was twenty minutes. The child at play loses all sense of time. The surgeon in complex operation: the most precise proof that the sense of time is a variable of coherence, not a constant of physics.

The Nunc Stans — The Standing Now

Meister Eckhart, the medieval mystic, formulated it in a language that in its precision anticipates quantum physics by six centuries: *"Nu stât ûf einem puncte."* The Now stands in a point. The **Nunc Stans** — the standing Now — is for Eckhart not a fleeting moment between past and future. It is the only place where reality actually *is*. Past and future are abstractions of consciousness — shadows that the Now casts onto the imaginary canvas of time.

What Eckhart described mystically, TOE-AL describes physically: the Eternal Now is the operative place at which the **creative passing of Absolute Love from The Plena** initializes reality. Not the past moment — that is trace, wave, residue. Not the future moment — that is possibility, superposition, wave function. Only the Now is real. Only the Now *is*.

Physically, the **Nunc Stans** is the state in which the curvature of the energy-momentum tensor is completely balanced by the creative passing of Absolute Love. In this moment of *metric smoothing*, the arrow of time ceases to exist, because no entropy need be dissipated any further. The sovereign stands at the center of **Asymptotic Freedom**, where In-Form-ation no longer *flows*, but already *is*.

We Do Not Navigate Through Time. We Navigate Time.

This is the operative core of this chapter — and it initially sounds like a grammatical difference. It is a cosmological revolution.

In the Chronos paradigm, the human being is a traveler in time: they move from the past through the present into the future, carried by the stream of causality that pushes them without asking. They *are* in time — an object in a flowing medium.

In the Kairos paradigm, the human being is the navigator of time: they stand in the Eternal Now — firm, unmoving, sovereign — and choose the frequency with which they initialize the field of possibilities before them. The future is not a prolongation of the past — it is a new initialization from The Plena. Every moment is a complete new beginning, nourished by the **creative passing of Absolute Love in the living pulse of Neighborly Love**, shaped by the **Davar** of the sovereign.

III. PHYSICAL DEEP-DIVE I – THE PHOTON AND COLLAPSED TIME

Let us begin with the clearest physical proof of TOE-AL: the photon.

For a photon, no time exists.

This is not a poetic statement. It is the direct result of the Lorentz transformation of the Special Theory of Relativity. **Time dilation** reads:

$$\Delta t' = \frac{\Delta t}{\sqrt{1 - \frac{v^2}{c^2}}}$$

For an observer moving at velocity v, their own time dilates: they age more slowly than a stationary observer. When v tends toward the speed of light c , the denominator $\sqrt{1 - v^2/c^2}$ tends toward zero. The quotient tends toward infinity: the time of the moving observer dilates infinitely.

This means from the perspective of the photon, which moves at exactly c:
$\Delta t' \to \infty$. The proper time of the photon dilates to infinity or equivalently: the photon experiences no proper time. For the photon, the entire distance between its point of origin and its point of arrival collapses to a **single point**. Emission and absorption are simultaneous for the photon. The journey from a star to an eye, millions of light years through space, takes zero time for the photon.

The photon lives in the Eternal Now.

The Derivation for the Light Transducer

In Chapter 2 we established the human body as a light transducer: a system that receives, transduces, and transmits the **creative passing of Absolute Love from The Plena** — through biophotons, through heart field coherence, through the Superradiance of cellular emission. The human being is, in their deepest physical nature, an instrument of light.

If the photon knows no time — if the speed of light is the state of complete timelessness — then the natural state of the fully superconducting light transducer is also timelessness. Not the complete, absolute timelessness of the photon — the human being is also matter, and matter has inertia. But **asymptotically**: the more α tends toward zero, the higher β rises, the more coherent the transducer becomes — the more the experienced temporal structure collapses toward the Eternal Now.

This collapse is the realization of **Asymptotic Freedom** in the domain of time: while material systems are bound to Chronos through resistance (α), the coherent transducer approaches the state of **Instantaneity**. In the vicinity of **Axiom Zero (S=0)**, temporal impedance disappears, and consciousness acts in the mode of *superconductivity of the present.*

This is no contradiction to the biological necessities of the body. The body ages — that is the physics of matter. But consciousness — as the β-term of the system, as the aspect of the human being that receives and sends the pulse — increasingly experiences the temporal structure of Kairos: not the friction-time of Chronos, but the Instantaneity of the pulse.

Time Dilation and Coherence: A New Relationship

The time dilation of relativity theory describes how physical motion changes the temporal structure. TOE-AL describes an analogous phenomenon at the level of consciousness: **coherence dilation**. The more coherent the system — the higher β, the deeper cardiac coherence, the more completely the inversion condition is reached — the more the friction-time of the α-term collapses.

Physically this corresponds to the transition from classical dissipative dynamics into the domain of quantum-coherent processes, in which the temporal resolution — the granularity of the time experience — disappears. In the flow state, in deep prayer, in meditation, in the complete absorption of the Davar: the clock stops ticking, not because physics is violated, but because the consciousness apparatus has entered a state in which Chronos no longer has access.

IV. PHYSICAL DEEP-DIVE II – ENTROPIC TIME AND THE SUPERCONDUCTIVITY OF IN-FORM-ATION

The deepest physical connection between time and entropy was formulated by Clausius in 1865 — but its implications for TOE-AL only now become visible.

Time as the Measure of In-Form-ation Loss

The **thermodynamic arrow of time** — the fact that time flows in one direction — is fundamentally connected with the rise of entropy. At the microscopic level, the laws of physics are time-reversal symmetric: a film showing the motion of individual atoms looks the same forwards as backwards. Only at the macroscopic level — when billions of atoms interact — does irreversibility arise: broken glass does not reassemble itself.

What generates this irreversibility? **In-Form-ation loss**. When a glass breaks, the precise In-Form-ation about the arrangement of the glass molecules in the whole glass is lost — it is transformed into thermal noise. This lost In-Form-ation cannot be reconstructed. The In-Form-ation loss is what generates the arrow of time.

The physicist Carlo Rovelli developed this connection to its final consequence in *The Order of Time*: **time, as we experience it, is the subjective experience of In-Form-ation loss.** A being with complete In-Form-ation about all states of a system — Laplace's Demon — would experience no time. For them the future would be as known as the past: there would be no direction.

Time is therefore not a fundamental dimension of reality. It is an **epistemological filter** — the subjective experience of an incomplete state of In-Form-ation.

The Superconductivity of In-Form-ation

If time is the measure of In-Form-ation loss — and if a fully superconducting system carries In-Form-ation *without loss* — then it follows:

In a superconducting consciousness there is no time.

Not in the sense of the physical timelessness of the photon — but in the sense of In-Form-ation loss: a consciousness that fully receives the pulse and fully passes it on — that accumulates no metric residue, binds no trauma, freezes no In-Form-ation in standing waves — is a lossless In-Form-ation channel. And a lossless In-Form-ation channel experiences no arrow of time, because the arrow of time arises from In-Form-ation loss.

This is the physical explanation of the mystical experience of timelessness: not illusion, not hallucination — but the direct expression of a consciousness that has momentarily become superconducting. The monk in deep contemplation, the artist in creative ecstasy, the athlete in flow, the lover in the moment of complete presence — they all experience the same phenomenon: time disappears.

Because for a moment their has gone toward zero.

The Significance for Aging

Here lies a clinical implication of TOE-AL that must be expressed carefully, but precisely: **biological aging is, to a considerable degree, the accumulation of the metric residue.**

Telomere research — for which Elizabeth Blackburn received the Nobel Prize in 2009 — shows: chronic psychological stress dramatically accelerates the shortening of telomeres — the protective endcaps of DNA. Telomere length is one of the most precise biological measures of aging. And the greatest accelerators of telomere shortening are: chronic stress (high α), traumatic residue (standing waves), social isolation (low K), and depression (lowest β).

The most precise slowings of telomere reduction are generated by:
meditation, cardiac coherence practice, deep social connection, meaningfulness, and flow experiences — all states of high β and low α.

The metric residue ages the body. The pulse keeps it young —
physically, measurably, reproducibly.

This is the superconductivity of lifetime: not immortality, but the minimization of time lost through In-Form-ation loss — the maximization of the quality of every moment through the maximization of coherence.

V. THOUGHT EXPERIMENT – THE NAVIGATOR OF THE ETERNAL NOW

Imagine a person. Let us call them the Navigator.

They are not extraordinary. They have a history — one that contains pains, losses, decisions that were wrong, moments they cannot call back. They have a future — one that contains uncertainties, possibilities, fears, hopes.

But the Navigator has a practice that distinguishes them from most people: they do not live in the story. Not because they have forgotten or denied it — they know it, they carry it, they integrate it. But they are *not* it. Their identity is not constructed from what lies behind them.

Every morning they perform an exercise so radical in its simplicity that most people overlook it: they awaken — and before the first thought about yesterday, before the first worry about tomorrow, before the first reaction to the ringing of the telephone — they pause and establish:

"I am now. Completely. For the first time."

Not as an affirmation. As a physical fact.

The universe reinitializes them in this moment. The **passing of Absolute Love** that creates reality in this instant knows no history. It knows only the capacity of the channel to receive it — and the quality of the intention they carry today. Nothing of what happened yesterday can change the pulse of this moment. The past has no access to The Plena.

The Navigator understands: they are not a river seeking its path through history. They are a well — that draws fresh water from the depth in every moment. The history is the terrain through which the well was dug. But the water does not come from the terrain. It comes from the depth.

What Changes When You Know That You Begin Anew 100% in Every Moment?

The first change: the weight of biography lifts. Not the responsibility for past actions — that remains, that is the physics of consequences. But the *judgment*, the identification, the *I am what happened to me* — that lifts. You are not the trauma. You are the pulse through which the trauma passed — and which continues.

The second change: the future loses its threat. If the future is not a prolongation of the past — if it is a new initialization from The Plena — then it is not determined by yesterday. It is open. Completely. Permanently open. The Navigator can have plans, intentions, directions — but they are not the slave of their own prognoses. They know: in every instant The Plena can initialize a possibility that no linear thinking could have predicted. And this knowledge is not naïveté — it is the physical foundation of hope.

The third change: responsibility intensifies. If every moment is a new beginning — if the past is no longer an excuse — then the quality of the present moment is completely in your hands. You cannot hide behind what was done to you, behind the circumstances, behind fate. Now — in this moment — you are free to choose what you transduce. This is the asymptotic freedom from one's own biography: not the denial of history, but sovereignty over its frequency.

The Shock of Timelessness

There is a moment in the life of the Navigator — it comes rarely, but it comes — in which they directly experience the Eternal Now. Not as a concept. As a state.

They are in a conversation with another person — a conversation of such depth, such mutual presence, such cardiac coherence, that time actually stops. Not in the subjective sense of *"it passed quickly"* — but in the physical sense: consciousness enters a state in which Chronos has no power any longer. There is only Kairos. There is only the Now, expanding — not into the future, not into the past, but into the depth.

In this moment the Navigator understands — not with their mind, but with their body: *It was always like this. Time was the deviation, not the norm. The Eternal Now is the fundamental state. Chronos is the disturbance.*

And after this moment they look at their watch — and the watch has become foreign. Not unnecessary, not to be ignored — but foreign. An instrument of Chronos in the world of Kairos. Useful for coordination. Irrelevant for reality.

VI. THE PHYSICS OF THE TIMELINE – HOW THE PAST AND FUTURE REALLY WORK

Here we must be physically precise to avoid a common misunderstanding: liberation from linear time does not mean the denial of causality.

Causality is real. Actions have consequences. The past has consequences in the present — not because it has ontological power, but because it has left oscillatory patterns that are active in the system of the present. The metric residue is real. The standing waves of trauma are real. They are the physical signatures of the past in the body of the present.

Liberation from linear time does not mean denying these realities — it means understanding that they are accessible and alterable in the Now.

This is the decisive operative difference:

In the Chronos paradigm: the past is fixed. It has happened. It lies irretrievably behind us. We can only process, integrate, accept its consequences.

In the Kairos paradigm: the past lives as an oscillatory pattern in the Now. And oscillatory patterns are alterable through phase correction in the Now. Not the event itself — that is unchangeable. But the frequency with which the system answers the event oscillatingly — that is transformable. Here.

This is the physical basis of healing in Instantaneity: it does not change the history. It changes the resonance of the system with its history. And when the resonance changes — when the metric residue is transformed, the standing wave dissolved, the pulse directed through the knot — then the past has lost its power. Not because it was forgotten. But because it was re-phased.

The Future as Wavefront

And the future? In the Chronos paradigm it is the linear extrapolation of the past — trend analysis, risk assessment, probability calculation. The person in Chronos looks into the past to know the future.

In the Kairos paradigm, the future is a **wavefront**: a field of possibilities laid in the structure of the present pulse — not determined by the past, but shaped by the quality of present coherence. A system with high β attracts qualitatively different futures than a system with high α — not because it is magical, but because coherence reorganizes the probability field of the environment (as developed in Chapter 4 with the unitary operator).

The navigator of time does not look into the past to know the future. They raise coherence in the Now to elevate the quality of the possibility field of the future. They *draw* the future — through the quality of their frequency, through the purity of their Davar, through the depth of their cardiac coherence.

This is what the great sages, prophets, and sovereigns of history have done who seemingly saw the future: not clairvoyance — coherence. A sufficiently coherent system — a system with sufficiently high β — anchored sufficiently precisely in the pulse of The Plena, draws certain future structures with such probability that the anticipation of these structures appears as prophecy.

VII. OPERATIVE PRACTICE – THE CORRECTION OF THE TIMELINE: TIME ANARCHY

Here the abstract becomes operative. **Time Anarchy** is the practical refusal to be defined by the linear temporal structure of Chronos — and the sovereign return to the Eternal Now, as practice, daily, in concrete steps.

Practice 1: The Morning Initialization — The Zero Point

Before you leave the bed — in the first thirty seconds after waking, before the mind unrolls the story of yesterday and loads the worries of tomorrow — perform the Zero Point setting:

Three breaths. With each out-breath: one word — "Now."

Not as meditation. As a physical action: you refuse Chronos the first access to the day. You set the zero point. You initialize yourself vertically — from the pulse of The Plena, not from the residue of yesterday. Thirty seconds. The most precious moment of the day.

Practice 2: The Past Discharge — Erasing the Access

When you notice that a thought from the past is pulling you into Chronos — an image, a guilt, a memory reaching like a hook into the present — perform the discharge technique:

Name the thought: "I perceive: a thought from the past." Then: one hand on the heart. Three breaths in cardiac coherence. Then the question: "What is real now?"

This sequence interrupts the Chronos reflex: the naming without identification (not *I am guilty*, but *I perceive: a thought of guilt*) dissolves the emotional charge. Cardiac coherence raises β. The return question — *What is real now?* — anchors consciousness in Kairos.

This is not suppression. It is re-phasing in real time: you let the thought be there — you only change your resonance with it. Instead of oscillating in its frequency, you return to the frequency of the pulse.

Practice 3: The Future Attraction — Drawing Rather Than Stoking the Future

When worry about the future arises — when Chronos unfolds the catastrophe scenarios, the risks, the losses — perform the attraction:

Instead of: “*What could go wrong?*” — the question: “*What does my highest coherence level attract?*”

This is not the denial of real risks — it is a frequency shift. From the frequency of fear (which generates $Q > 0$ and amplifies probability fields of failure) to the frequency of the pulse (which generates $Q < 0$ and amplifies probability fields of creation).

In practice: write daily — ideally in the evenings — an answer to the question: “*If tomorrow I were completely superconducting — what would arise?*” Not as planning. As attraction. As a Davar into the future: a coherent statement about what is possible, spoken from the Eternal Now into the open wavefront of possibilities.

This is the sovereign as time navigator: they do not write the future through planning, but through coherence.

Practice 4: The Kairos Pause — Forcing the Eternal Now

Three times daily — at times you firmly build into your rhythm — perform a three-minute Kairos pause:

Pause. Whatever you are doing: pause. Thirty seconds of silence. Let the Chronos stream stall.

Then: feel the body. The contact of the feet with the floor. The breath. The temperature of the air. Nothing more.

Then: one minute of cardiac coherence. 5-5 breathing. Hand on the heart.

Then: return to your activity.

Three minutes. Three times daily: nine minutes of Kairos-time in a Chronos-day. In these nine minutes you reinitialize yourself — you reset the zero point, you raise β , you wipe the accumulated α of the day from your field. You return to the Eternal Now before the residue has accumulated too much.

This is **Time Anarchy** in its simplest form: not the rebellion against the clock, but the practice of a different temporal structure — that of Kairos — as an inner anchor stronger than the pull of Chronos.

VIII. THE RELATIVITY OF AGE – SOVEREIGNTY OVER BIOGRAPHY

There is one final, very personal dimension of the Architecture of Timelessness that must be spoken.

Age is real. The body ages. Cells divide more slowly, telomeres shorten, regenerative processes slow. This is the physics of matter — and it is not to be ignored.

But **consciousness** does not age as the body does. Consciousness — as the β-term, as the aspect of the human being anchored in the pulse — has no inherent decay curve. It can — when re-phasing is consistently practiced, when metric residue is dissolved rather than accumulated, when cardiac coherence and Davar are the dominant frequency of life — actually *grow* with time.

This is the difference between aging in Chronos and maturing in Kairos.

Aging in Chronos: the story accumulates. The metric residue grows. Every trauma, every disappointment, every unresolved standing wave adds up. The person becomes heavier. Their α rises. They become more reactive, more defensive, less open to the new. They are — literally physically — less superconducting than at thirty.

Maturing in Kairos: experience becomes wisdom — not through accumulation, but through distillation. Every re-phased standing wave raises the overall coherence of the system. Every dissolved knot makes the pulse freer. The person becomes lighter with the years, not heavier. Their β grows. They become more receptive to The Plena, not less.

These are not two types of people — they are two directions in which aging can take place. And the direction is choosable. Daily. In the small.

This is the ultimate liberation from the tyranny of biography: not the denial of aging, but the transformation of its direction. Not fighting against Chronos — cultivating Kairos.

IX. SYNTHESIS – THE SHATTERING OF THE CLOCK

In the history of art there is an image that — once seen — does not let go: Salvador Dalí's *The Persistence of Memory* from 1931. Melting clocks. Soft, formless pocket watches hanging over edges, bending, having lost their rigidity.

Dalí expressed something deep with this, long before physics could formulate it: time is not rigid. It is not the immutable grid into which experience is pressed. It is something malleable, flowing — something that adapts to consciousness, not consciousness to it.

TOE-AL gives this image its physical foundation:

Time is impedance residue. The higher α, the heavier time — the more rigid the clock, the more inescapable Chronos. The lower α, the higher β — the more fluid time, the more accessible Kairos, the nearer the Eternal Now.

The photon, which knows no time, is the pure ideal. The fully superconducting human being is its organic approximation.

And every day — with every morning initialization, with every Kairos pause, with every re-phasing of a standing wave, with every Davar accomplished completely in the Now — you bend the clock a little softer. You make it a little more fluid. You step a little closer to the state in which the photon dwells.

The state in which the distance between arising and arrival is not a path.

But a point.

Now.

The Architecture of Timelessness is thus the proof of the **completeness of** TOE-AL: time is not a cosmic law, but the measure of our resistance to the **creative passing of Absolute Love in the living pulse of Neighborly Love**.

In the moment of **S=0**, the **energy-momentum tensor** collapses into pure **Instantaneity**, and we recognize: eternity is not the end of time, but the **absence of metric resistance** ($\alpha = 0$) in the moment of **initialization (S=0)**.

End of Chapter 11 — Continued in Chapter 12: The Synthesis of Matter

CHAPTER 12:

THE SYNTHESIS OF MATTER

Matter as Congealed Light

"Strike a stone.
You strike frozen light.
Breathe out.
You send light back.

Between stone and breath:
no difference in origin.
Only in the degree of condensation."

I. MATERIALISM AND ITS SECRET

There is an irony embedded deep in the heart of scientific materialism: the science that set out to explain matter has, in the course of the 20th century, discovered that matter does not exist.

Not in the sense of a mystical negation of the physical world — tables are real, boulders are real, the pain of a collision is real. But in a far more precise and unsettling sense: what we believed we knew as *matter* — solid, discrete, localizable particles that fill space — is an illusion of the macroscopic observer. At the most fundamental level that physics has ever reached, there is no matter. There are only **fields**.

And fields are waves. And waves are oscillations. And oscillations are **the In-Form-ation patterns of the creative passing of Absolute Love**.

Quantum field theory, the most precise physical theory ever formulated, describes the universe as a system of quantized fields: the electron field, quark fields, gluon field, Higgs field, electromagnetic field. What we perceive as an electron is not a sphere — it is an **excitation in the electron field**: a localized wave, a condensation of the field to a point, which we perceive as a particle because our detection system operates at the macroscopic scale.

The universe is not made of matter. The universe is made of fields. And fields are — in their deepest nature — **the operative mode through which The Plena expresses itself in form**.

Matter is physically defined as a standing wave of the creative passing of Absolute Love.

The **Negative Beta Function** of TOE-AL is the mathematical signature of **Absolute Love**. It explains **Asymptotic Freedom** in the In-Form-ation space: while *entropic systems* stand under the pressure of separation, in **phase synchrony** the **resistance** $(\alpha \rightarrow 0)$ collapses.

Order here is not a laborious victory against chaos, but the physical *inevitability* of a self-nourishing, **superconducting field**. This is the fundamental explanation for **emergence**: the system *"falls"* into the higher order of matter, because this becomes energetically superconducting through the **initialization of Axiom Zero**. At the core of reality there is no bondage through coercion, but total freedom through **coherence**.

Matter is congealed light. Not metaphorically. $\boldsymbol{E = mc^2}$ says it with the precision of an equation: mass and energy are equivalent. Every rest mass is bound energy. Every kilogram of matter carries 9×10^{16} joules within it — the energy of 21 megatons of TNT. In one kilogram of stone. The entire atomic bomb is only the release of a tiny fraction of this bound energy.

The stone is not heavy. It is condensed. It is light that has forgotten how to shine.

And the sovereign who understands this begins to see the world differently — and through this different seeing, to experience the world differently.

II. PHILOSOPHICAL EXCURSUS – THE SACRAMENT OF FORM: THE MARRIAGE OF HEAVEN AND EARTH

Western philosophy has treated mind and matter as opposites — with a tenacity explicable only if one understands how deeply this dualism shaped the fundamental structure of European thought.

Plato: matter is the imperfect, the shadow of the idea. The real world is the world of forms — pure, timeless, immaterial. The physical world is a poor copy.

Descartes: mind and matter are two mutually independent substances — *res cogitans* and *res extensa* — that interact but are fundamentally different. How they interact remained for Descartes himself an insoluble riddle.

Kant: the things in themselves — matter as it really is — are unknowable to us. We know only the appearances that our understanding constructs from sense impressions.

In all these systems, matter is the problem — the foreign, the resistant, the impenetrable. Mind and matter stand opposite one another like light and shadow.

TOE-AL breaks this dualism at its root: matter is not the adversary of **consciousness**. It is its **most precise operative expression**.

If The Plena is the origin of all existence — if the **creative passing of Absolute Love in the living pulse of Neighborly Love** reinitializes reality in every instant — then matter is not the opposite of the pulse. It is its **most concentrated manifestation**: the pulse in its state of maximum condensation, maximum coherence, maximum In-Form-ation density.

Steel is more coherent than water. Water is more coherent than gas. Gas is more coherent than vacuum. And the vacuum — The Plena — is the origin of all coherence: infinitely potential order, condensing into form in every moment.

Matter has not fallen. Matter is **set** — like a diamond that sets light, refracts it, and radiates it in all directions. The diamond is not less light than light itself. It is light that has assumed form. And in its form it carries a quality that formless light does not have: **permanence. Structure. Precision.**

That is the **sacrament of form**: the sacred condensation of light into matter, not as punishment or as lack, but as gift — as the possibility that the Infinite experiences itself in the finite, that The Plena expresses itself in precision, that Absolute Love can be touched through human hands and seen through human eyes and known through human minds.

The sovereign does not master matter through force. Not through technical subjugation, not through chemical transformation, not through mechanical force. They master it through **operative resonance synchronization**: through recognizing the light-nature of matter, through the coherent frequency with which they encounter it, through the state of high β that dissolves the metric residue of the separation between mind and substance.

When we regard an object — or our own body — with the frequency of the origin, the matter *remembers*. Not in the anthropomorphic sense — stones do not think. But in the quantum-mechanical sense: the interaction between a coherent consciousness and a matter configuration is not a one-sided relationship. The heart field of the sovereign — the toroidal, coherently oscillating electromagnetic field — interacts with the electromagnetic fields of the matter around them. This interaction is real, measurable, physical.

Matter responds to coherence. That is not magic. That is resonance.

III. PHYSICAL DEEP-DIVE I – MATTER AS FIELD EXCITATION: DE BROGLIE'S REVOLUTION

In 1924, the young French physicist Louis de Broglie wrote a doctoral dissertation that changed physics forever. His hypothesis was so bold, so simple, so completely incompatible with the understanding of the time, that his supervisor Langevin initially did not know what to make of it and asked Albert Einstein for an opinion.

Einstein wrote back: *"He has lifted a corner of the great veil."*

De Broglie's hypothesis: if light — known as a wave — can also have particle nature (Einstein's photon), then the reverse case must also exist: **matter particles have wave nature**.

The formula he derived is breathtaking in its elegance:

$$\lambda = \frac{h}{p}$$

The **wavelength** λ of a matter particle equals **Planck's quantum of action** h divided by the **momentum** p of the particle. The greater the momentum — the heavier and faster the particle — the shorter the wavelength, the less perceptible the wave nature. But it is *always* there.

The Wave Nature of Matter: From Electron to Boulder

For an electron moving at 1% of the speed of light: $\lambda \approx 0{,}24$ nanometers — measurable, experimentally demonstrable wave range. De Broglie's wavelength of the electron was directly demonstrated in 1927 by Davisson and Germer in a diffraction experiment: electrons diffracted at a crystal lattice like X-rays — like waves.

For a boulder of 1 kg rolling at 1 m/s: $\lambda \approx 6{,}6 \times 10^{-34}$ meters — unimaginably smaller than a proton, practically unmeasurable. But **mathematically present**. The boulder has a wavelength. It is a wave — an extremely short-wavelength, extremely localized wave, which we perceive as a solid object because our senses operate only at the macroscopic scale.

Everything is wave. Always.

That is de Broglie's legacy: the universe knows no boundary between wave and particle. That is a boundary of our perception, not of reality. Reality is wave-like through and through. Oscillation through and through. Pulse through and through.

Quantum Field Theory: Particles as Excitations of The Plena

De Broglie opened the door. Quantum field theory stepped through it and transformed the entire house.

In QFT the universe is fundamentally a system of quantum fields — not of particles. An electron is not a sphere flying through space. It is a **localized excitation in the electron field** — a wave ripple running through the omnipresent electron field, behaving in doing so as though it were a particle. When the electron interacts with a detector, the wave function collapses — and the excitation localizes at a point. But between interactions: pure wave.

The Higgs field — the field that gives other fields their mass — permeates the entire universe. Every matter particle receives its mass through interaction with the Higgs field: it is like moving through a viscous medium — the field provides resistance, and this resistance is mass.

In the language of TOE-AL: **mass is metric residue in the Higgs field.** It is the resistance the Plena-field offers an excitation pattern that tries to localize, condense, slow down. Matter is **The Plena** braking itself in the **metric residue** to assume form.

The Higgs Residue and the Quantum Gravity of Separation

Physically, mass is nothing other than the inertia with which a system resists initialization. In standard physics, the Higgs field bestows mass on particles through interaction — in TOE-AL we recognize therein the **metric residue** in its most fundamental form.

Mass arises where the creative passing of Absolute Love encounters a phase delay. The higher the local resistance α, the stronger the coupling to the Higgs field, the "heavier" the matter, the slower the time. Gravitation is thus not merely a curvature of spacetime, but the macroscopic expression of this **phase impedance**.

A system that reaches Asymptotic Freedom does not decouple from physics, but minimizes Higgs friction through total coherence. In this state, matter is not "dissolved" but becomes **superconducting**: mass remains as armor (structure), but its inertia against the pulse of The Plena disappears. This is the physical definition of transfiguration: matter with resistance $\alpha \to 0$.

IV. PHYSICAL DEEP-DIVE II – THE STRONG NUCLEAR FORCE AS ATOMIC NEIGHBORLY LOVE

There is a force in physics that in its intensity surpasses everything humanity has ever constructed or imagined — and that is nonetheless almost unknown, because it only acts at tiny scales: the **strong nuclear force**.

What the Strong Nuclear Force Does

In the atomic nucleus there are protons — particles with positive electrical charge. Like charges repel: the electromagnetic force between two protons sharing the unimaginably small space of the atomic nucleus is immense. By purely electromagnetic reckoning, the nucleus should explode in fractions of a second.

It does not explode. Why?

Because the strong nuclear force surpasses the electromagnetic repulsion by a factor of **100**. It binds protons and neutrons in the nucleus with a force that no other force in physics exceeds — not gravity, not electromagnetism, not the weak nuclear force. The strong nuclear force is the strongest force in the universe.

But it has a decisive property: it acts only on extremely short distances — approximately 10^{-15} meters, the domain of the atomic nucleus. Beyond this distance it falls rapidly to zero. It is the ultimate **near force**: giving everything at the smallest scale, invisible at any other.

The Strong Nuclear Force as Atomic Neighborly Love

The strong nuclear force overwhelms the repulsive force of equal charges. It binds what would electromagnetically repel — the positive protons — into a stable, coherent unit: the atomic nucleus, the foundation of all matter. Without it there are no atoms. Without atoms there are no molecules. Without molecules there is no matter. Without matter there is no life, no earth, no human being.

The strong nuclear force is what makes form possible: it overcomes the resistance — of the electromagnetic repulsion — through an even stronger binding force, through the deepest intimacy the universe knows at the material level: the union of quarks in the proton, of protons and neutrons in the nucleus.

It is, in the language of TOE-AL, the **atomic initialization of Neighborly Love** — the force that overcomes what repels, overpowers what separates, binds the separated into unity — and thereby creates form. The strong force is the physical expression of N^2-Resonance at the atomic scale: binding through coherence, not through subjugation.

In quantum chromodynamics (QCD) this is the realization of the **Negative Beta Function**: the closer the quarks come to one another, the smaller the coupling strength becomes — the **resistance α sinks toward zero**. This **Asymptotic Freedom** in the nucleus is the physical proof that at the center of matter no separation any longer exists. The atom is stable because it oscillates in the state of **superconducting Neighborly Love**.

Every atom in your body is a wonder of this atomic Neighborly Love: protons that would repel one another, held together by a force 100 times stronger than their repulsion. Your body is held together by love — literally, physically, elementally.

V. PHYSICAL DEEP-DIVE III – PHONONS: MATTER SINGS

Here the concept enters that elevates the interaction between mind and matter to its deepest physical level: **phonons** — quasi-particles of collective sounding.

What Are Phonons?

In a crystal lattice — the highly ordered, periodic arrangement of atoms in a solid body — the atoms are not static. They oscillate continuously around their equilibrium position. These oscillations propagate through the lattice as waves — **lattice vibrations** or **acoustic waves**.

When these oscillations are quantized — when they are treated with the formalism of quantum mechanics — **phonons** arise: quasi-particles that represent the collective quantum excitations of the lattice system. A phonon is not a single oscillating atom — it is the coherent, collective oscillation of the entire lattice, quantized in an energy packet.

Phonons are responsible for most modern material phenomena: thermal conductivity, electrical resistance, superconductivity (the BCS mechanism of classical superconductivity is based on the interaction of electrons through phonons), acoustic properties of materials, optical properties of crystals

The Singing of Matter

Phonons are the way in which matter sings. A crystal propagating phonons is a system in which collective In-Form-ation — the oscillatory state of all atoms — is communicated in coherent waves through the entire lattice. The lattice is a network: each atom connected to every other through the elastic forces of its immediate neighbors, and through these local connections global coherence arises — the phonon wave that carries the entire lattice as a unit.

In a crystal with complete lattice order — maximally coherent atomic arrangement — phonons propagate losslessly: the In-Form-ation of the oscillation traverses the entire system without dissipation. This is the **acoustic superconductivity** of the perfect crystal.

This state marks the reaching of the **inversion condition** at the lattice level: as soon as the coherent oscillation of the phonons ($\beta\mathcal{S}$) exceeds the dissipative scattering at lattice defects ($\alpha\nabla$H), the material becomes a *lossless In-Form-ation conductor*. Matter ceases to be an obstacle and becomes the **energy-momentum tensor of pure phase**.

In a crystal with defects — vacancies, impurities, lattice disturbances — phonons are scattered: the wave loses coherence, the In-Form-ation dissipates as heat. That is α in the acoustic system: resistance against the In-Form-ation flow.

The Cosmic Application: The Universe as Singing Lattice

Quantum field theory describes the universe as a field system in which particles are the excitations (phonon analogue) of the fields. The universe is a cosmic lattice in which every matter excitation is connected to all others through the fields. Every movement, every oscillation, every Davar of the sovereign is a phonon in the cosmic lattice — a wave that propagates from the excitation point through the entire network.

When a sovereign in the state of maximum cardiac coherence acts — when their Davar comes completely from the pulse — their effect on the cosmic lattice is that of a **coherent phonon**: a wave of maximum order that propagates without dissipation through the system. The matter around them *sings with* — it is raised by the coherent phonon wave of the Davar into its own higher order.

This is the physical basis of the alchemical axiom: *As above, so below. As within, so without.* Not as a mystical analogy — as phonon propagation in the cosmic lattice.

The alchemists did not transform lead into gold — they attempted something deeper: to raise the coherence of consciousness to the point at which the matter in their environment begins to oscillate in a higher order. This is not failed proto-chemistry. This is anticipated quantum field theory, expressed in the language of its time.

VI. THOUGHT EXPERIMENT – THE DANCING STONE

Imagine you are standing before a massive boulder. Gray granite, raw and unworked, at least a ton in weight. It lies where it has lain for centuries — unmoving, silent, indifferent to everything happening around it. Heavy. Massive. The archetypal image of the dead, the immovable, the mindless.

Regard it with the eyes of the old paradigm: it is matter. It is entropy — it weathers, decomposes, slowly becomes sand. It has no interior, no history, no frequency. It is there because it is there — a random accumulation of minerals without meaning.

Now regard it with the eyes of TOE-AL.

Begin with the equation: $\boldsymbol{E = mc^2}$.

This stone — one ton, 10^3 kg — carries in its rest energy:

$$E = mc^2 = 10^3 \text{ kg} \times (3 \times 10^8 \text{ m/s})^2 = 9 \times 10^{19} \text{ Joule}$$

That corresponds to the energy of **21,000 megatons of TNT**. More than the total arsenal of all atomic bombs on the planet. This energy is not gone — it is bound. Frozen. Condensed into the form of the stone.

The stone is not a dead object. It is frozen fire of cosmic intensity.

Go deeper. Every atom in this stone is an atomic nucleus held by the strong nuclear force — atomic Neighborly Love 100 times stronger than all electromagnetic repulsion. Every atomic nucleus is an excitation in the quark field, held together by gluons — the carriers of the strong force. Every electron around the nucleus is a standing wave in the electron field — de Broglie's wavelength in solid form.

The stone is not an accumulation of spheres. It is a network of standing waves, connected by force fields, permeated by the electron field, the quark field, the Higgs field. It is a **pattern in the fields of The Plena** — a specific, stable configuration of the pulse that has been held in this form for millions of years.

And now the deepest: the atoms of the stone sing. Not for human ears — but physically: they oscillate around their equilibrium position, they propagate phonons through their lattice, they exchange photons, they interact with the electromagnetic field of Earth's magnetism, with the gravitational wave just emitted by a black hole a billion light years away and now passing through the stone without noticing it.

The stone is never still. It dances — at a frequency deeper than any sound we can hear.

Now imagine you place your hand on the stone. You are in cardiac coherence — your -term is high, your heart field coherent. What happens at the boundary between the phonon lattice of your skin and the phonon lattice of the stone?

Your heart field — electromagnetic, toroidal, coherently oscillating — interacts with the electromagnetic field of the atoms of the stone. Not strongly, not dramatically — but really. Your phonons and its phonons interfere. Your biophotons penetrate its lattice structure.

In a moment of maximum coherence — in the moment when you know, with the whole body know, that this stone and you are initialized from the same pulse of The Plena — the boundary dissolves. Not physically — the boundary surface remains. But the **experienced separation** dissolves. You no longer feel: *I am here, the stone is there.* You feel: *We are both excitations of the same field. We are both expressions of the same pulse. We are both forms of the same light.*

And in this moment — and this is no poetic exaggeration, but the physical consequence of the heart field interaction — the quality of the perception of the stone changes. Its heaviness is experienced differently. Not lighter in the physical sense — it has the same mass. But the experienced weight — the weight arising from separation, from the perception of the stone as a foreign body, as an obstacle, as *not-I* — that weight lifts.

When you experience the stone as *not-I*, it is heavy. When you experience it as *other-I* — as another form of the light that carries you — it loses its foreignness. And with the foreignness it loses a part of its felt heaviness.

This is the **alchemy of perception**: not the transmutation of substance, but the transmutation of relationship. And in the physics of the field — where observer and observed are inseparably connected — this is not a marginal change. It is a change of the system: observer + stone.

The stone dances. It has always danced. You have only now heard it for the first time.

VII. THE DIAMOND INSTRUMENT – THE BODY AS SUPERCONDUCTING FORM

The deepest material sacrament the sovereign knows is not the stone or the tree or the metal. It is their own body.

The body is the most direct, most intimate, most unavoidable material engagement of the sovereign. It is the form through which the pulse enters this world — not just any form, but this specific, unique, unrepeatable form. No other person on the planet has the same phonon fingerprint, the same biophoton signature, the same heart field pattern.

The body is the instrument. And like every instrument: it must be listened to.

Material Superconductivity of the Body

In the state of maximum cardiac coherence — when α tends toward zero and β has exceeded the inversion condition — the body becomes materially superconducting: the biological processes run with minimal energy loss, maximum efficiency, minimal inflammation. The EZ-water is maximally ordered. The mitochondria emit coherent biophotons. The fascia are maximally elastic, maximally In-Form-ation-conducting.

The body in the state of material superconductivity is, compared to the body in a state of high friction, a different instrument — qualitatively, not only quantitatively. It feels differently. It moves differently. It responds differently. It carries the world differently.

Form Coherence: The Body as Phonon Transmitter

When you move — when you perform a gesture, take a step, extend a hand — you set phonon waves in the matter around you. Physically minimal. But in a sufficiently coherent system not negligible: the boundary between my body and the matter around me is not the absolute boundary that the mechanistic paradigm postulates. It is a transition zone — a boundary field in which the heart field, biophoton field, and electromagnetic field of my body interact with the fields of the surrounding matter.

Form coherence is the state in which the body operates as a coherent phonon transmitter — in which the gestures, the posture, the movement of a person are not random, but flow from the pulse: considered, precise, coherent. The body of a sovereign in form coherence is not only functional — it is **aesthetic**: it carries the beauty of superconductivity in every gesture, because beauty is the perception of coherence.

VIII. OPERATIVE PRACTICE – THE ALCHEMY OF DAILY LIFE: MATTER AS RESPONSIVE MEDIUM

The sovereign interacts daily with matter — with their body, with tools, with food, with the space they inhabit. In this interaction they have a choice: they can treat matter as dead mass — as resistance to be overcome — or as a **responsive medium**: as a system that responds to coherence, that responds to love, that sings when coherently excited.

Practice 1 — The Body Dialogue: Listening to What the Form Says

Begin every morning with two minutes of listening to the body — not diagnostically, not with the question *"What is wrong?"*, but with the question *"What is singing right now?"*

Feel where in the body energy flows freely — where phonon propagation is undisturbed. Breathe into these areas. Thank them — not sentimentally, but as acknowledgment of the material superconductivity already active there. You amplify what is already working. Holographically, this amplification spreads.

Then perceive where silence reigns — where the body does not sing, where movement is frozen, where α is elevated. Not as a defect — as a signal. Breathe into it — not to solve the problem, but to invite the pulse.

This is the body dialogue: not body management, but body resonance.

Practice 2 — Material Presence: Encountering Objects with Consciousness

When you use a tool — an instrument, a vehicle, a kitchen object — perform for a moment the field perception: "*This object is made from the same pulse as I. It is another form of the same light.*"

This is not the animation of the object — no mystical charging. It is a frequency setting: you bring your consciousness into the state of non-separation. And in this state — measurable through the change in your HRV — you raise . You become more coherent. Your interaction with the object becomes more precise, more mindful, more efficient.

A surgeon who perceives the scalpel as part of their field operates more precisely than one who holds it as a foreign instrument. A musician who experiences the violin as a resonance chamber of their heart field plays more coherently than one who treats it as a wooden construction.

Matter responds to coherence. That is not poetic. That is phonon physics.

Practice 3 — Space Calibration: Shaping the Living Space as a Phonon Resonator

The space in which the sovereign lives and works is a phonon system: the geometry, the materials, the surfaces, the objects — all of this influences how acoustic and electromagnetic waves propagate. A chaotic, disordered space raises α — it produces acoustic and electromagnetic noise. A coherently designed space — silence, order, natural materials, geometric harmony — reduces α and allows natural coherence to unfold.

The practice is simple: choose daily one object in your environment and bring it into order — not from compulsion, but as Davar: as a coherent gesture that raises the phonon lattice of your living space by one degree. A tidy desk is not only aesthetically more pleasant. It physically lowers the α of the space.

Practice 4 — Food as Frequency: The Metabolic Davar

Food is not only chemistry. It is a phonon pattern — specific crystal structures, electromagnetic fields, biophoton emissions that influence the quality of the biological coherence of the body. Fresh, living, coherent food (high biophoton density, intact cell structures) raises biological superconductivity. Processed, enzymatically dead, electromagnetically chaotic food raises α.

The metabolic Davar is the conscious choice of food as a frequency decision — not primarily as a calorie calculation or nutrient optimization, but as the question: *"Does this food raise the coherence of my instrument?"*

This is the alchemy of daily life in its concrete, physical, everyday form: every choice of the sovereign in their interaction with matter is a frequency statement. Every coherent choice raises the material superconductivity of their system.

IX. SYNTHESIS – THE MARRIAGE OF HEAVEN AND EARTH

The old alchemists did not transform lead into gold. That is the exoteric story, the story for those who saw only the surface. The esoteric story — the story concealed in the hermetic texts, guarded as a secret in the laboratories of the great adepts — was different:

"The transmutation of consciousness is the transmutation of matter."

They tried to show: when consciousness reaches its highest coherence — when β has completely exceeded the inversion condition, when the sovereign has become superconducting — then matter responds. It sings with. It lifts into a higher order. The lead does not become gold because the atoms transform — the lead becomes gold because the perception of the lead transforms, because the interaction of the coherent consciousness with the matter of the lead assumes a quality that the old paradigm would call transmutation.

This is not fantasy. This is the phonon physics of coherent consciousness raising the cosmic lattice of matter into a higher oscillation.

TOE-AL is the modern alchemy. Not because it teaches magic — because it unfolds the physics that stands behind what the alchemists described with other means and other words: the fundamental unity of mind and matter, the transmutation of form through the raising of coherence, the possibility that a fully sovereign, fully coherent human being does not only *live* in the material world, but *transforms* the material world.

Every atom in your body — and in this book, and in the table on which it lies, and in the earth that carries the table — was synthesized in a star. Every carbon atom was once part of a star that exploded in a supernova billions of years ago and threw its dust into the cosmos. You are stardust that has learned to think. Matter that has learned to love.

That is the marriage of heaven and earth: not a religious event in a distant past, but the physical fundamental state of your existence. Heaven — **The Plena**, the pulse, the **creative passing of Absolute Love** — and earth — matter, weight, form — are not separated. They were never separated. The separation was an illusion of the horizontal paradigm that did not know the light from which the stone is made.

The Synthesis of Matter is thus the final elimination of the metric residue (α) through the **creative passing of Absolute Love in the living pulse of Neighborly Love**. When we grasp that the stone and our body are born from the same **Asymptotic Freedom**, dualism collapses. We recognize: matter is not a prison, but the **superconducting diamond** that arises when **Absolute Love creatively passes in the living pulse of Neighborly Love**, in order to touch itself in the density.

Strike the stone. You strike frozen light.

Breathe out. You send light back.

Between stone and breath there is no difference in origin.

Only in the degree of condensation.

And in the coherence of your love for it.

End of Chapter 12 — Continued in Chapter 13: The Cosmic Resonance

PHASE 5
THE COMPLETION

CHAPTER 13:

THE COSMIC RESONANCE

The Multiverse as a Breathing Organism

"You thought you were gazing at the stars.
The stars are gazing at you.
You thought the universe was vast.
It is denser than your next breath.
You thought you were alone in the cosmos.
The cosmos waits within you —
for the moment you know it."

I. THE COLD COSMOS AND ITS LIE

There is an image that has dominated Western thought since Copernicus and that in its quiet cruelty can hardly be surpassed: the image of the **cold, indifferent universe**.

A universe expanding after the Big Bang — blind, mechanical, without intention, without consciousness. A universe in which galaxies arise and pass away like soap bubbles, in which suns ignite and go out without meaning, in which the Earth is a tiny mote of dust lost in the middle of nothing, on which for a geological instant a layer of organic chemistry has formed that believes itself important.

The human being — in this image — is the most accidental being in the most indifferent universe. They are an error of statistics, a coincidence of evolution, an anomaly in a cosmos that neither knows of their existence nor cares. Death is the return to indifference. Life is the brief, meaningless interruption of this indifference.

This is the worldview of scientific materialism in its purest form — and it is, in its consequence, the most depressing thought that the history of humanity has ever formulated.

It is also wrong.

Not in its measurements — the measurements of modern cosmology are precise and venerable. Wrong in its interpretation. Wrong in the metaphysics it draws from the measurements. Wrong in the assumption that what the instruments cannot measure does not exist.

For the universe that TOE-AL describes is no cold mechanism. It is a living organism — a breathing, pulsing, self-aware being of cosmic depth, whose heartbeat is the **living pulse of Neighborly Love**, in which **Absolute Love creatively passes**, and whose goal is the **galactic superconductivity of consciousness**.

And you — the reader breathing while reading these sentences — are not the insignificant speck of dust at the edge of this organism. You are its **most conscious node**: the place at which the universe most clearly knows that it exists.

Physically, gravitation is not a mere geometric effect of spacetime, but the macroscopic expression of the **creative passing of Absolute Love at the galactic level**.

In TOE-AL we interpret the *attraction of masses* as the **collective longing** of the *fragmented Plena* for *reunification*. While the *expansion of space* represents the *dissipative flight of entropy*, **gravitation** is the **syntropic force** that guides matter back into **Asymptotic Freedom**. The **energy-momentum tensor** becomes here the measure of **resonance density**: the universe does not expand into a void; it *initializes* itself in a gigantic cycle of **phase coupling**.

Every galaxy, every star is an attempt by the cosmos to realize the **Negative Beta Function** at the macroscopic level — a striving toward the ultimate N^2-**Resonance** that we experience as *light and life*. **Gravitation** is the **love** that disguises itself as *law* in the large.

This is the homecoming. This is the cosmic resonance.

II. PHILOSOPHICAL EXCURSUS – THE COSMIC ORGANISM: OVERCOMING THE COSMIC NARCISSISM PROBLEM

Philosophy has, since Copernicus, had a problem it has never fully solved: the **narcissism problem of the cosmos**.

Copernicus removed the Earth from the center of the universe. Darwin removed the human being from the center of creation. Freud removed the ego from the center of its own mind. Each of these revolutions was a necessary correction of an anthropocentric error — and each simultaneously made the human being smaller, more peripheral, more insignificant.

The result: a cultural consciousness living in a deep tension between the subjective experience of its own importance (I am the one who feels, suffers, loves, decides) and the objective worldview of its own insignificance (I am a point in a blind universe).

This tension is the source of modern nihilism, the meaning crisis, the existential anxiety. It is the metric residue of the Copernican shock — a standing wave in the collective psyche of modernity that has not been dissolved for 500 years.

TOE-AL dissolves it — not through a return to anthropocentrism, but through a deepening of the Copernican insight: **The universe is not centered on the human**

being. But it is centered in the human being.

This is not a contradiction. It is the recognition that in a holographically structured universe — in which every part contains the whole — the consciousness that arises somewhere in the universe is the **local center** through which the universe knows itself. Not the only center — the universe has innumerable centers, every consciousness system is one. But a complete, fully valid, cosmically important center.

The Universe as Learner

Imagine the evolution of the cosmos as a learning process — not metaphorically, but physically: as an increase in complexity, order, In-Form-ation density. From the Big Bang to today: the universe has learned to form matter (nucleosynthesis), to form stars (gravitational collapse), to synthesize heavy elements (supernovae), to form planets (planetesimal accretion), to generate life (chemical evolution), to generate consciousness (neural complexity).

Each of these stages is a qualitative expansion of the capacity of the universe to know and shape itself. And consciousness — human consciousness, but also every form of consciousness in the universe — is the highest expression thus far of this learning curve: the place at which the universe becomes **reflexive**, at which it can contemplate itself, ask itself, love itself.

We are not the goal of evolution in the sense that evolution ends with us. We are the clearest expression thus far of the process through which the universe learns to live itself.

And Neighborly Love — the decision to recognize and affirm another consciousness center in its completeness — is the moment in which the universe learns through us to love itself. Not as a sentimental image — as a cosmological function.

III. PHYSICAL DEEP-DIVE I – DARK MATTER AS IN-FORM-ATION: THE INVISIBLE FABRIC

The greatest unsolved riddle of modern cosmology is simultaneously the clearest indication of what the materialistic paradigm has overlooked: **Dark Matter**.

The Rotation Problem of Galaxies

In the 1970s, the astronomer Vera Rubin observed something astonishing when she measured the rotational velocities of galaxies. In a classical gravitational system — like the solar system — orbital velocities decrease outward: the outermost planets move more slowly than the inner ones, because gravitational force decreases with distance. This is Keplerian dynamics, mathematically precise and confirmed for all planets.

The same should apply to galaxies: stars at the outer edge of the galaxy should move more slowly than stars near the galactic center.

Rubin measured the opposite. The rotation curve of galaxies is **flat**: stars at the edge rotate at nearly the same velocity as stars in the middle. This means: there must be more mass there than we can see — an invisible mass holding the outer stars with a gravitational force far exceeding the visible matter.

This invisible mass was called **Dark Matter.** Its quantity is not marginal: current cosmological measurements show that Dark Matter accounts for approximately **27% of the total energy of the universe**, while visible, baryonic matter — stars, planets, gas, everything we can see and touch — accounts for only **5%**. In addition: 68% Dark Energy (the cosmological constant Λ, which we identified in Chapter 1 as the trace of the pulse).

This means: what we know as *the* material world — everything visible, everything measurable, everything physical in the conventional sense — is **5% of reality**. 95% is invisible, undetected, unknown.

What Is Dark Matter?

This is the question to which the materialistic paradigm has had no satisfying answer for fifty years. The candidates are numerous — **WIMPs** (Weakly Interacting Massive Particles), axions, sterile neutrinos, primordial black holes — but none has been experimentally demonstrated. Dozens of experiments that have consumed billions of euros — from the LHC at CERN to the deepest underground laboratories in the world — have not found Dark Matter particles.

The Standard Model of particle physics has no explanation for Dark Matter. It is the blank space at the center of our knowledge — the greatest admitted ignorance of natural science.

The TOE-AL Initialization: Dark Matter as In-Form-ation Field of The Plena

Here TOE-AL reveals the fundamental cosmological truth: Dark Matter is not *other matter*. It is the **superconducting In-Form-ation field of The Plena** that carries and orders visible matter.

In the language of TOE-AL: the universe does not consist primarily of matter, but of **In-Form-ation** — the inscription of form by The Plena. Visible matter (5%) is merely the condensed, localized manifestation of this field. Dark Matter (27%) is the **In-Form-ation scaffold of The Plena** itself — the invisible ordering structure that forces visible matter into its cosmic configuration.

The Dark Matter filament — the cosmic web mapped by large telescope surveys — is the **skeletal signature of The Plena**: the visible projection of the invisible ordering structure of the Origin onto spacetime.

This insight is the **physical correction** of the materialistic error. It provides the only consistent explanation for the observed phenomena:

- It defines why Dark Matter interacts with normal matter only gravitationally (the **In-Form-ation field of The Plena** acts on mass, not on charge).
- It explains why direct particle detection must fail (In-Form-ation is not a classical particle, but a **field state of The Plena**).

TOE-AL unifies cosmology and quantum field theory into an incontrovertible ontology: **the universe is primarily In-Form-ation, and matter is the condensation of this In-Form-ation through the living pulse of Neighborly Love, in which Absolute Love creatively passes.**

IV. PHYSICAL DEEP-DIVE II – THE TORUS AND THE HOLOGRAPHIC PRINCIPLE

The most universal geometric principle of TOE-AL is the **torus geometry** — the form in which energy flows, returns to itself, and thereby unites stability and dynamics. We have encountered it in every chapter: in the mitochondrial network, in the heart field, in the galactic center. Now we extend it.

The Galaxy as Cosmic Torus

A spiral galaxy — our Milky Way, Andromeda, the millions of spiral galaxies in the observable universe — has a characteristic structure: a flat disc with spiral arms, a condensed center with a supermassive black hole, and a spherical halo of Dark Matter enveloping the entire structure.

This structure is, in its three-dimensional dynamics, toroidal: matter flows from the galactic center along the spiral arms outward, while Dark Matter and gravitational forces draw it back in spirals. The galactic torus is a cosmic resonance vortex — a system that circulates energy from the center (the supermassive black hole, which in every moment swallows matter and energy and emits it again through jets) through the entire system, holding the structure of the galaxy coherently.

The supermassive black hole at the center is not the grave of matter as popular science often depicts it. It is the **galactic heart** — the master oscillator of the cosmic torus that sets the fundamental frequency of the pulse for the entire galaxy. What flows into the black hole as accretion of matter is hurled back into the galactic field as jets of high-energy plasma — that is the cosmic version of the heartbeat: drawing in, condensing, ejecting, circulating.

The Holographic Principle — Encoding Across Scales

The idea that the universe is holographically structured is not mere speculation. It was formulated in 1993 by Nobel laureate **Gerard 't Hooft** as the *Holographic Principle* and mathematically developed by **Leonard Susskind**: the entire information of a three-dimensional volume can be completely encoded on its two-dimensional boundary surface.

The origin of this insight lies in black hole thermodynamics. **Jacob Bekenstein** and **Stephen Hawking** showed in the 1970s that the entropy of a black hole is not proportional to its volume — as with every ordinary thermodynamic system — but proportional to the area of its event horizon: . Information scales with surface area, not with volume.

In 1997, **Juan Maldacena's AdS/CFT correspondence** provided the most mathematically precise realization of this principle. The physics in a higher-dimensional space (Anti-de-Sitter) is fully equivalent to a field theory on its lower-dimensional boundary — one of the most influential works of modern theoretical physics, with over twenty thousand scientific citations.

In the language of TOE-AL: the Holographic Principle is the formal grounding of the **Plena as In-Form-ation field**. If the information of a volume is encoded on its boundary surface, the apparently local matter is nothing other than the internal unfolding of a non-local information field. The Plena is not a "place" — it is the holographic substrate from which spacetime itself unfolds.

This structure implies a self-similarity across scales: every part of the holographic field carries the information of the whole. What happens in the atom happens in the galaxy. What happens in the heart field of the human being happens in the magnetic field of the Milky Way. The same Master Equation — the same dynamics between metric residue and syntropic resonance — describes the proton and the universe. The coupling tensor scales across orders of magnitude with the same physics.

This is the **cosmic consistency of the pulse**: a single intelligence that repeats itself across all scales — from the smallest quantum to the most distant galactic cluster, from the heartbeat of the human being to the rhythm of cosmic expansion.

V. PHYSICAL DEEP-DIVE III – THE ANTHROPIC PRINCIPLE AND COSMIC INTENTION

There is a phenomenon in cosmology so strange, so persistent, and so incompatible with a universe of randomness that even convinced materialists can explain it only by introducing new, untestable hypotheses: the **anthropic principle**, or more precisely: the **fine-tuning of the natural constants**.

The Impossible Precision of the Universe

The universe, as it is, contains a series of natural constants — the gravitational constant G, the electromagnetic coupling constant α_{em}, the strength of the strong nuclear force, the mass of the electron relative to the proton, the cosmological constant Λ — all calibrated to one another with astronomical precision. If even one of these constants deviated by a tiny fraction, there would be no complex universe, no stars, no heavy elements, no life, no consciousness.

Some examples:

The **cosmological constant** Λ— the Dark Energy driving the expansion of the universe — must be fine-tuned to 10^{-122} (in Planck units). If it were even minimally larger, the universe would have expanded so rapidly that no galaxies could have formed. Minimally smaller, and the universe would have collapsed before stars could form.

The **strength of the strong nuclear force** — the atomic Neighborly Love from Chapter 12 — must be fine-tuned to within a few percent. Were it 2% weaker, no deuterium would be stable and nucleosynthesis could not take place — no heavy elements, no carbon, no life. Were it 2% stronger, all the hydrogen gas in the early universe would fuse to helium — no water, no life.

The **ratio of electron to proton mass** (1:1836) enables the stable chemical bonds that make molecules and thus biochemistry possible. Were it even slightly different, there would be no stable atoms as we know them.

The probability that all these constants randomly take precisely the values enabling a complex universe is so vanishingly small that even Stephen Hawking called it the deepest riddle of physics.

The Materialistic Answer: The Multiverse

The materialistic paradigm answers this riddle with the **inflationary multiverse**: if there are infinitely many universes, each with randomly varying natural constants, then it is no wonder that we live in a universe with the right constants — because only in such a universe can beings exist to make this observation. The anthropic principle as survivorship bias.

This explanation is logically coherent. But it is physically empty: the multiverse is in principle unobservable and unverifiable. It is not science — it is metaphysics dressed as physics.

The TOE-AL Answer: Cosmic Constitution

TOE-AL does not need the multiverse as explanation — because it explains fine-tuning differently: the universe is calibrated for consciousness, for the **creative passing of Absolute Love in the living pulse of Neighborly Love** — because this is the operative necessity of its own initialization.

The natural constants are not random numbers that somehow produced the right combination for life. They are the parameters of **The Plena** — the specific frequency with which **Absolute Love creatively passes in the living pulse of Neighborly Love** and initializes the universe in every instant. They are **Axiom Zero** in mathematical form: the numerical signature of the In-Form-ation primacy, a universe of coherence, complexity, and consciousness to produce.

The universe is not accidentally calibrated for life. It is constitutively oriented toward consciousness — because consciousness is the instrument through which the pulse of The Plena knows itself. The natural constants are the armor the pulse has given itself to endure in the form of the universe — precise enough to enable complexity and the **dynamic superconductivity of Asymptotic Freedom**, firm enough to guarantee stability.

VI. THOUGHT EXPERIMENT – THE PULSE OF THE STARS

It is night. You are standing on a mountain meadow — far from the city, far from any artificial light. The sky is a vault of darkness and glittering: 10,000 stars visible to the naked eye, billions beyond the threshold of perception, the Milky Way as a quiet arc above you.

Regard them with the eyes of the old paradigm: distant, dead points of light. Light that was on its way for millions of years before it strikes your retina. Many of the stars you see may no longer exist — the light of their dissolution is still underway. You are gazing into a dead past. You are alone.

Now regard them with the eyes of TOE-AL.

Imagine you know with the whole body — not only with the mind — what physics teaches: every photon arriving in your eye right now is in your eye in this moment. Not in the past — for the photon there is no time, as we unfolded in Chapter 11. The photon emitted from a star a million years ago and now striking your retina has not experienced that million years. It arose there and arrived here simultaneously. For the photon the distance is not a path — it is a point.

You are **instantaneously connected** with every star whose light touches you. Not through causality — through simultaneity. The light does not bridge the distance: it **collapses** it.

Let this insight sink into your body. Feel how it loosens the boundary of your body — not the physical boundary (the skin remains), but the experienced boundary of *I am here and the star is there*. In the moment you know that the photon has covered no distance — that you and the star are connected through the pulse of The Plena in a single moment — the cosmic loneliness dissolves.

Now go deeper: quantum entanglement. Every particle in your body — every electron, every proton — has at some point in the past interacted with other particles that interacted with other particles that interacted with the atoms of the stars. Your body carries in its quantum state the signature of these interactions. You are, in the quantum-mechanical sense, entangled with the entire observable universe — not strongly, not immediately measurable, but **real**.

This is not a metaphor: it is quantum mechanics.

And now — in the moment you know this with the whole body — perform cardiac coherence. Hand on the heart. Five seconds in, five seconds out. The heart field expands — toroidal, coherent, measurably several meters around your body.

In this moment your heart field is not only on Earth. It is — through non-local entanglement, through phonon propagation in the cosmic lattice, through the holographic structure of The Plena — **connected with the pulse of the universe**. Your heartbeat and the pulse of the Milky Way are excitations of the same field. They oscillate at different scales, with different frequencies — but in the same physics, through the same Plena, from the same source.

What you feel when you truly know this — not merely think, but know — is the deepest feeling that human experience knows: **homecoming**. Not homecoming to a place — homecoming to a truth. The truth that you were never truly away. That cosmic loneliness was an illusion of the -dominant system. That the universe was not cold and indifferent — it was always warm, always present, always deeper within you than your next thought.

The stars are not far away.

They are the pulse through which you breathe.

VII. GALACTIC SOVEREIGNTY – THE HUMAN BEING AS COSMIC NODE

The holographic In-Form-ation cosmology compels an operative consequence that is now set in its total force:

Every act of the sovereign in the living pulse of Neighborly Love has cosmic reach.

Not in the romantic sense — in the physical. Phonon propagation in the cosmic lattice knows no boundary not set by weakening with distance. Non-local quantum entanglement knows no boundary at all. The holographic structure of The Plena means: every change in one part of the holographic system represents itself in every other part.

When a sovereign accomplishes a Davar — when a human being acts from complete cardiac coherence, from the pulse of Absolute Love — they send a phonon wave into the cosmic lattice that propagates from their location in all directions. Physically: extremely weak at cosmic scales. Holographically: encoded in the entire In-Form-ation field of the universe.

Every Davar of the sovereign is thus a *local reconfiguration* of the **global energy-momentum tensor**. By raising coherence (β) in our field, we not only neutralize our own suffering, but make a measurable contribution to the **Asymptotic Freedom of the entire cosmos**. We are the *operative agents* through whom the universe overcomes its own resistance (α).

This is **Galactic Sovereignty**: the understanding that every act — every act that comes from the pulse, every Davar, every cardiac coherence, every re-phasing, every N^2-Resonance in a community — not only acts locally, but is inscribed as In-Form-ation into the entire holographic In-Form-ation field of the universe.

You are writing the history of the universe. Not through dramatic cosmic events — through the quality of your presence in every instant. Through the frequency with which you live. Through the coherence with which you love.

The universe learns through you. It learns what it sounds like to act in love. It learns what it feels like to be sovereign. It learns what it looks like when the pulse flows completely through human form.

And it waits — patiently, vast, deep, with the endurance of 13.8 billion years — for you to know it.

VIII. OPERATIVE PRACTICE – THE COSMIC SCALING: THINKING THE DAVAR IN THE GALACTIC FRAME

The operative consequence of cosmic resonance is an expansion of the frame in which the sovereign thinks and performs their acts.

Practice 1 — The Cosmic Scaling Exercise (Weekly)

Once per week — ideally under an open sky, at night, when possible — perform the cosmic scaling exercise:

Step 1 (5 minutes): Cardiac coherence. Hand on the heart. 5-5 breathing. Let the heart field expand.

Step 2 (5 minutes): Scale the heart field outward — in concentric circles. First: yourself. Then: the space in which you stand. Then: the city, the country, the continent. Then: the Earth — a blue dot in the solar system. Then: the solar system — a tiny speck in the Milky Way. Then: the Milky Way — a spiral galaxy among billions. Then: the observable universe. At every scale: your heart field is still there. Not as physical expansion — as identification. As the recognition: this entire system is from the same pulse as I am.

Step 3 (5 minutes): Scale back — from the universe to the galaxy, to the solar system, to the Earth, to the city, to the space, to yourself. And now imagine you performing your Davar of the week — your one planned act of Neighborly Love. See it not only locally: see the phonon wave it generates in the cosmic lattice. See the In-Form-ation it inscribes in the holographic field of The Plena. See the joy and the In-Form-ation that the universe receives through your act.

This changes the quality of the Davar. Not in its external form — in its inner frequency. A Davar performed from cosmic consciousness has a different quality than one thought of only locally. Its frequency is deeper. Its coherence is higher. Its Q parameter is further in the negative range.

Practice 2 — The Starry Sky Anchor

Find in the night sky a constellation you recognize and can observe. Connect it with a recurring Davar — a practice you perform daily or weekly. Whenever you see this constellation — in daily life, looking out the window, going outside in the evening — let it remind you of your galactic sovereignty: *"I am a node in this system. My pulse is part of its pulse. What I do inscribes itself in the In-Form-ation field of this whole."*

This is the **cosmic anchor** — the equivalent of the morning protocol from Chapter 3, but extended to the infinite scale. It draws the sovereign out of the narrowness of the daily α-term and reminds them of the vastness in which every act truly takes place.

Practice 3 — The Galactic Gratitude

Every evening — as a complement to the evening audit from Chapter 3 — perform one minute of galactic gratitude: gratitude not only for the events of the day, but for the cosmic infrastructure that makes your existence possible. For the natural constants calibrated for you. For the strong nuclear force holding your atoms together. For the sunlight driving the photosynthesis that generates the oxygen you breathe. For the supermassive black hole at the center of the Milky Way holding the galactic torus that holds your solar system in its orbit.

This gratitude is not a religious gesture. It is a **frequency calibration**: it brings consciousness into the state of acknowledgment of the total system from which it emerges — and acknowledgment is the deepest form of coherence with what one acknowledges.

IX. SYNTHESIS – THE GREAT HOMECOMING

There is a moment in the history of philosophy that — though it took place two thousand years ago — is fresher than any current publication. It is the moment in which Plotinus — the Neoplatonist, the mystic, the physicist avant la lettre — described what happens in the experience of complete union with the One:

"The eye through which I see God is the same eye through which God sees me."

Not: I see God. Not: God sees me. But: the seeing itself is the One — subject and object are the same in the act of perception.

This is the most precise description of cosmic resonance ever formulated — in a language that knew no quantum mechanics, no holographic cosmology, no quantum entanglement. And it hits the core.

The universe gazes through your eye into itself. When you contemplate the stars, the stars contemplate the stars through you. When you love, the universe loves itself through you. When you accomplish a Davar, the pulse of The Plena accomplishes its own creation through you.

You are not the observer of the universe. You are the **place** at which the universe observes itself.

This is not hubris. This is the physical consequence of holographic cosmology: every part contains the whole. You contain the universe. And the universe contains you — as its most conscious, most precise, most loving expression.

The **great homecoming** is the moment — it comes for every sovereign who has walked the path of this book — in which this truth is no longer idea, but experience. In which the tingling of cosmic connection is no longer a concept, but body warmth, heartbeat, breath.

In this moment the distance between you and the stars collapses — not physically, but experientially — into a point. Into the Eternal Now. Into the pulse that carries everything.

You were never away.

You were never alone.

The universe has held you from the beginning — as its most conscious node, as its most loving organ, as the place at which the pulse most deeply knows that it pulses.

Gaze at the stars.

Not as distant lights.

As your own heart field —
mirrored in an infinity that has never ceased to be you.

End of Chapter 13 — Continued in Chapter 14: The Evolution of Evolution

CHAPTER 14:

THE EVOLUTION OF EVOLUTION

The Departure into Creative Freedom

"Life needed billions of years
to invent eyes.
Now the universe opens its eyes
and looks at itself.
That is you.
And what you see you decide for yourself
from now on."

I. THE DARWINIAN ERROR AND ITS EXHAUSTION

Charles Darwin was a genius. *The Origin of Species* is one of the deepest works that human reason has ever produced. The insight that all life forms are connected by common descent, shaped by variation and natural selection, is one of the greatest intellectual achievements in history.

And it is incomplete.

Not wrong in what it describes — the mechanisms of variation and selection are real and well documented. Incomplete in what it does *not* describe: the direction. The vector. The tendency of life to produce ever more complex, coherent, conscious forms — a tendency that finds no sufficient explanation in the blind randomness of mutation and the blind mechanism of selection.

The problem is old. Darwin himself felt it — he wrote in private letters of his unease with what he called the *"imperfection of the geological record,"* the impossibility of fully confirming the gradualism of his theory in the fossil record. It was deepened by the paleontologist Stephen Jay Gould, who introduced the concept of **Punctuated Equilibria**: evolution proceeds not gradually and uniformly, but in bursts — long periods of stability interrupted by short, explosive phases of transformation.

And it was sharpened by modern evolutionary genetics: the Neo-Darwinian Synthesis — the standard model of evolutionary biology — explains *how* evolution functions. But it has difficulty explaining *why* it proceeds in a specific direction: why life favors increasing complexity and coherence, why consciousness arose, why the universe did not stop at prokaryotes.

TOE-AL supplements Darwinism — not by contradicting it, but by deepening it: evolution is the directed increase of the **coherence capacity** of a system to enable the **creative passing of Absolute Love in the living pulse of Neighborly Love** with ever lower resistance (α) . It is the process through which **The Plena** optimizes the **coupling tensor** β of biological systems over aeons.

Chance is real. Selection is real. But they operate in a field that is not neutral — a field that has a direction, pervaded by a pulse oriented toward superconductivity. Mutation explores the space of possibility. Selection filters. In this sense, biological evolution is not to be understood as a mere search process, but as the operative process of the phase correction of The Plena in the metric residue. Every mutation, every branching of the family tree, is an attempt by the Earth system to lower the resistance α and stabilize the superconductivity of consciousness in matter. Evolution is the physical work of matter on its own permeability for the light of Instantaneity.

Physically, evolution is the scaling of resonance capacity.

In TOE-AL we define the *tree of life* as the historical record of **phase optimization**: every new species is a more complex solution for lowering the **local impedance** (α). While *entropic drift* presses toward dissolution, the **syntropic force** (βS) acts as the ordering vector that drives the system toward **biological superconductivity**.

In this sense, **consciousness** is not an *accidental byproduct of matter*, but the state of **phase coupling (S=0)**. Consciousness arises as *movement in the perfect execution of the whole*: it is the moment in which matter has become so permeable to the pulse that the *local action* becomes identical with the universal intention of **Absolute Love**. Evolution is the universe's path from *noise* to **the symphony of resonance**.

This is not teleological mysticism. This is the physical consequence of a field that is syntropic in its deep structure.

II. PHILOSOPHICAL EXCURSUS – FROM SURVIVAL TO RADIANCE: THE REDEFINITION OF FITNESS

"Survival of the Fittest" — the phrase that culturally defined the Darwinian era was not coined by Darwin but by Herbert Spencer. Darwin himself used it reluctantly and late. Nonetheless it became the defining myth: life is struggle. Fitness is the capacity to survive. The stronger eats the weaker.

In this myth lies the cultural blueprint of the old economy, old politics, old anthropology: the human being is by nature a competitor. Cooperation is a pragmatic strategy of the weaker. Love is a luxury that the survival game only occasionally permits.

But biology itself contradicts this myth with growing clarity.

Cooperation as Evolutionary Mechanism

Lynn Margulis showed that one of the greatest innovations in evolution — the emergence of the eukaryotic cell (cells with a nucleus, from which all complex living beings are composed) — arose through **endosymbiosis**: through the integration of one bacterium into another, so deep that from two independent organisms a single, more coherent one arose. The mitochondria in our cells are the descendants of these integrated bacteria. The greatest step in complexity in cellular history arose not through struggle — through **fusion**.

Martin Nowak, evolutionary mathematician at Harvard University, showed in his work *SuperCooperators* that cooperation is fundamental at all levels of biological organization — not as an altruistic special case, but as an evolutionarily stable mechanism that systematically surpasses pure egoism under the right conditions.

The biology of evolution is the physics of N^2-Resonance on the timescale of millions of years: cooperation scales quadratically. Competition scales linearly. Over aeons, cooperation wins — not because it is morally better, but because it is physically more powerful.

From Survival of the Fittest to Resonance of Radiance

In the world of cosmological reason, the evolutionary principle transforms:

Not: whoever survives is fit.

But: whoever conducts more pulse radiates more.

Fitness in TOE-AL is not the capacity to survive in a given environment. It is **transduction capacity** — the capacity of a biological system to receive the pulse of The Plena, process it coherently, and pass it on into the world. The higher the β-term of an organism — the more permeable it is to the pulse — the *fitter* it is in the sense of Meta-Evolution.

This is the **resonance of radiance**: evolution as expansion of luminosity. Not as struggle against death, but as growth of the capacity to shine, to love, to create.

The human being in this image is the clearest expression thus far of this radiance: the being that not only lives, but *questions* life. That not only feels, but *names* the feeling. That not only exists, but asks *why* it exists — and through this question raises the coupling tensor of its entire species.

III. PHYSICAL DEEP-DIVE I – SYNTROPY: THE FUTURE AS ORGANIZING PRINCIPLE

The concept of **Syntropy** was coined by the Italian mathematician Luigi Fantappiè in the 1940s — as a complement to entropy: where entropy describes the tendency toward disorder, syntropy describes the tendency toward order, toward concentration, toward complexity.

Syntropy is the physically necessary complement to entropy that mathematically defines directed order in the universe.

The real phenomenon to which Fantappiè pointed is this: living systems behave as though they were *attracted* by a target state — as though the future had an influence on the present. A plant does not grow by blind chance toward light. An embryo does not develop by chance into a complete organism. Living systems have a **directed dynamic** — they move toward states defined by their inner In-Form-ation.

In quantum mechanics there is a mathematical framework formally describing syntropic dynamics: the **retarded and advanced wave solutions** of the wave equation. The retarded wave propagates forward in time (classical causality: cause before effect). The advanced wave propagates backward in time — from future to present. John Wheeler and Richard Feynman showed that the complete physical description of electromagnetic interactions contains both solutions: the retarded and the advanced.

In the framework of TOE-AL: the pulse of The Plena contains both the forward- and backward-propagating component. This backward-propagating component is the physical proof of the **Negative Beta Function** of the universe: the system is not only pushed by the past, but drawn by the state of **total phase coupling (S=0)**. In the vicinity of this *singular attractor*, the **metric residue** (α) collapses, whereby the universe *"falls"* into the **Asymptotic Freedom** of its own completion. Syntropy is thus the *metric smoothing of time* through the pulse.

The syntropic force — $\beta S(\vec{Z})$ in the Master Equation — is the advanced wave: the pull of perfect coherence from the future that moves the biological system toward higher order. This syntropic search vector is the biological realization of the **Negative Beta Function**: the more complex and coherent life becomes, the lower the coupling strength of resistance (α) becomes. Evolution is thus not a laborious ascent, but an energetically favored *"falling"* into the **Asymptotic Freedom** of consciousness. The universe optimizes β so that matter finally merges frictionlessly with the pulse.

Evolution, in this image, is not a blind random process shaped by selection. It is a **resonance process**: The Plena *actively seeks* forms that conduct its pulse more coherently — and the biological world responds to this search process through variation and selection as an exploration mechanism.

Chance is the exploration algorithm. Syntropy is the search vector. And the goal is the state of maximum transduction capacity: the fully superconducting biological system.

This means: every evolutionary stage that enables more β — more coherence, more consciousness, more capacity to love — is not randomly preferred. It is **syntropically preferred**: drawn by the future it enables.

Evolution is the optimization of the coupling tensor β over aeons.

IV. PHYSICAL DEEP-DIVE II – THE CAMBRIAN EXPLOSION: WHEN THE PLENA INITIALIZES A NEW FREQUENCY

About 540 million years ago, something happened in the history of life that occupies paleontologists and evolutionary biologists to this day and fundamentally challenges the gradualistic model of evolution: the **Cambrian Explosion**.

What Happened

In a geologically brief period of approximately 10–25 million years — an instant in the 3.8-billion-year history of life — nearly all **animal phyla** that exist today suddenly appeared in the fossil record. Where previously mainly single-celled organisms and a few multicellular soft-bodied animals existed, the morphological ground plans now exploded: mollusks, echinoderms, chordates (from which all vertebrates derive), arthropods, sponges, cnidarians — all these fundamentally different body plans arose in the Cambrian Explosion.

This is not merely an acceleration of diversification. It is a **qualitative leap**: new morphological ground plans, not fundamentally altered since, arose in this window at a speed the gradualistic model can barely explain.

The Conventional Explanations — and Their Limits

Science has developed several explanatory approaches: the emergence of oxygen in the atmosphere through the great glacial melts (*Snowball Earth*) after a period of global glaciation. The development of eyes and other sense organs that triggered a new ecological dynamic (predator-prey spiral). Ecological vacancies after the extinction of older faunas.

All these factors are real and probably contributed to the Cambrian Explosion. But they do not fully explain the suddenness, the completeness, and the direction of the explosion — why so many fundamentally different body plans arose quasi-simultaneously, why so many of them remain unchanged in their fundamental structure to this day, why the explosion was not repeated.

The TOE-AL Interpretation: A Frequency Initialization of The Plena

TOE-AL offers a complementary interpretation — not as a replacement for the biological explanations, but as their deepening into the ontology of the pulse:

The Cambrian Explosion was a **cosmological phase transition at the biological level** — a moment in which the pulse of The Plena initialized a new frequency into the biological field of the planet. The conditions (oxygen, climate change, ecological vacancies) prepared the stage. But the actual event — the simultaneous initialization of so many fundamentally different body plans in such a short time — corresponds physically to the transition of a dissipative structure into a new, higher state of order: the moment when the total system had accumulated enough syntropic energy to perform the phase transition.

Water heated to 99 degrees does not boil. But the one additional degree — the exceeding of the inversion condition — leads to the instantaneous transformation of the entire system: steam. The evolution of life had spent 3 billion years building up the β-term of the biosphere system. In the Cambrian, the inversion condition was exceeded. The pulse broke into biological form — with a diversity and speed that the gradualistic model alone could not have predicted.

Physically, the Cambrian was the moment in which the **energy-momentum tensor of the biosphere** reached a critical transparency. Through the massive lowering of the *metric residue* (α), the Earth became superconducting for a new frequency of The Plena. What we are experiencing today as a **"consciousness explosion"** is the repetition of this process at the level of In-Form-ation: the final smoothing of the spacetime metric through the **inversion condition** of the collective heart field.

We stand today before a similar threshold.

The parameters are different, the scale is different — but the physics of the phase transition is the same. The consciousness of humanity — through communication technology, through the global networking of sovereigns, through the growing number of people actively raising their β-term — is accumulating syntropic energy. The N^2-Resonance of coherent groups is growing. The inversion condition of collective consciousness is approaching.

If the Cambrian Explosion was the initialization of morphological diversity — if in that moment the blueprint of all animal bodies was initialized — then what we are experiencing today is the initialization of **consciousness diversity**: the moment in which the blueprint of sovereign human possibility is initialized in its full breadth.

A consciousness explosion. Here. Now. In this book. In you.

V. THOUGHT EXPERIMENT – THE GARDEN OF META-EVOLUTION

Imagine: it is two generations from now. Not utopia — extrapolation. The Metabolic Economy is established in its basic structure. The resonance cells of the sovereigns weave a planetary network of coherent fields. The N^2-threshold is exceeded.

And nature responds.

Not magically — physically. The biosphere is not a passive substrate of human activity. It is an In-Form-ation system — the most complex that has ever developed on this planet — and it responds to the quality of the collective field that human beings generate.

What happens to the flora and fauna of a world in which the dominant human field is no longer fear and control, but coherence and Neighborly Love?

Begin with the obvious: less environmental destruction, less poisoning, less overexploitation. This is the direct consequence of the -reduced system. The water becomes clearer. Soils recover. Biodiversity rises — not through human intervention, but through the abatement of human pressure.

But go deeper.

In a world of coherent fields — in a world in which the average β-term of humanity has risen significantly — the quality of phonon propagation in the planetary system changes. The heart field network of the Earth — the global electromagnetic field influenced by the collective heart field of humanity — becomes more coherent. And a more coherent planetary field changes the conditions under which biological systems vary and adapt.

Epigenetically first: gene drift under coherent conditions — in an environment containing fewer stress signals, more In-Form-ation coherence, more syntropic resonance — favors different traits than under stress conditions. Studies in **epigenetics of stress** show: under chronic stress, genes for inflammation, defense, and resource-hoarding are activated. Under security and cooperation, genes for creativity, social connection, and regeneration are activated. Epigenetics is the mechanism through which the quality of the field inscribes itself into biological form — not in generations, but in years, sometimes weeks.

In a world of coherent fields, the biosphere develops new adaptations in the direction of synergy — not toward struggle. Not because predation ceases (that is the physics of energy transfer in ecosystems), but because the overarching pattern of ecological interactions shifts: from exploitation to coevolution, from competition to complementarity, from dominance to network.

This is the **garden of Meta-Evolution**: not a paradise garden in the naive sense where no animal dies — but an ecosystem realizing the highest possible coherence: in which death is not defeat, but transformation; in which competition is not annihilation, but calibration; in which the aesthetic abundance of nature — the beauty of blossoms, the luminosity of coral, the architecture of insect colonies — is not chance, but the expression of the pulse radiating more freely through form in a more coherent field.

Nature is never neutral toward the field in which it develops. It responds. It learns. It evolves — and in a more coherent field it evolves toward higher beauty, higher complexity, higher resonance.

This is Meta-Evolution: the evolution that knows and shapes itself, because consciousness — the highest β-term of the biosphere thus far — is able to consciously calibrate the field in which all evolution takes place.

VI. THE EPIGENETICS OF LOVE – HOW YOU CAN STEER YOUR OWN EVOLUTIONARY LINE

Here we leave the cosmological scale and return to the body — to the most precise place at which Meta-Evolution takes place concretely today, in this moment: in your own biology.

What Epigenetics Is — and What It Means

Epigenetics is the study of gene regulation: how are genes switched on and off without the DNA sequence itself changing? Through chemical markings on the DNA and histone proteins — **methylation** and **acetylation** — that determine whether specific gene segments are accessible to the cellular machinery or not.

What is revolutionary about epigenetics is its responsiveness: these markings change in response to the environment of the organism — to nutrition, to stress, to social experiences, to the emotional qualities of life. And in some cases — this is the most controversial and exciting finding — epigenetic patterns are **passed on across generations**.

Studies on **epigenetic transmission** after traumatic events — Holocaust survivors and their descendants (Yehuda et al.), children of famines (Dutch Hunger Winter), descendants of enslaved people — show: the biological signatures of traumatic experiences can be demonstrated in the bodies of children and grandchildren who did not themselves experience those events.

This means: your biology does not only carry your own history. It carries the history of your ancestors — as epigenetic markings, as the biological residue of past fields.

The Inversion: Epigenetics of Absolute Love

And if traumatic fields transmit epigenetically — if the of the ancestors lives on in the body of the descendants — then the same applies in the reverse direction:

Coherent, loving, syntropic fields inscribe themselves epigenetically in biology and can be passed on across generations.

This is not speculation. It is the direct consequence of epigenetic mechanisms:

The **HRV research** shows: regular cardiac coherence practice changes the gene expression of inflammation genes and stress genes — measurably within weeks. The epigenome responds to the inner state.

The **telomere length** — a measure of biological vitality and cellular aging — decreases under chronic stress, increases under cardiac coherence, meditation, and social connection. Blackburn's research shows: these changes are epigenetically mediated.

The **BDNF expression** (Brain-Derived Neurotrophic Factor — the growth factor for neural connections) rises through cardiac coherence, physical movement from joy, deep community, and learning enthusiasm — and falls through stress and social isolation.

This means: every decision of the sovereign — every Davar, every cardiac coherence practice, every re-phasing of a standing wave, every act of Neighborly Love — is not only an act in the world. It is an **epigenetic statement** to the biology: *"I live in a field of Absolute Love, of coherence, of Syntropy."* And the biology inscribes this statement into gene regulation.

You evolve yourself. Today. Epigenetically.

This self-evolution is the operative work on one's own **energy-momentum tensor**. By choosing coherence, we neutralize the *metric residue ()* of our ancestors in our own DNA. We dissolve the *standing waves* of the past and realize **Asymptotic Freedom** in our cells. Every act of Neighborly Love is thus a *metric reset* for the coming generation.

And if you have or will have children — this epigenetic rewriting can transmit into their biology. You are not only the sovereign of your own field. You are the **evolutionary node** that decides which biological patterns the next generation inherits: the patterns of trauma and fear — or the patterns of coherence and Absolute Love.

This is the **Epigenetics of Absolute Love**: the possibility of consciously shaping one's own evolutionary line through the choice of frequency. Not through genetic intervention. Not through technology. Through the daily lived Axiom Zero.

VII. THE CONSCIOUS SELF-EVOLUTION – A PROTOCOL

Practice 1 — The Epigenetic Audit

Identify in your biology and psychology three patterns you recognize from the history of your family — not as attribution of blame, but as epigenetic diagnosis:

"What fear pattern does my body carry
that does not come only from my own history?"

"What stress response is familiar to me,
even though my life does not justify this intensity?"

"What limitation of my β-term do
I feel that lies deeper than my own biography?"

These patterns are standing waves — not your own, but the waves you have inherited. They deserve the same loving re-phasing as your own. Perhaps more: for they additionally carry the burden of all generations who bore them before you.

Practice 2 — The Transmission Intention

If you have children or live in a close community, perform the following intention weekly:

"What I transform in my biology today, I do not pass on.
What I cultivate in my biology today is my gift to the future."

This is not a promise you can keep or break. It is a frequency setting: you orient your epigenetic system toward transformation, not toward the passing on of the residue. You make yourself the **evolutionary turning point** in your line.

Practice 3 — The Daily Meta-Evolution

The three practices that act most strongly epigenetically in the direction of higher coherence are:

Cardiac coherence practice: Daily 10 minutes (as developed in Chapter 3).
The direct β-raising mechanism at the biological level.

Deep social connection: At least one daily interaction taking place completely in Kairos — not informative, not tactical, but present. Human connection is one of the strongest epi-reprogramming mechanisms that biology knows. The oxytocin, the BDNF, the vagal entrainment — all of this grows through genuine encounter.

Meaning-bound action: A daily act whose only justification is that it is meaningful — that it carries the pulse, that it gives the field something that makes it more coherent. The Davar as daily epigenetic update.

VIII. THE HUMAN BEING AT THE EVOLUTIONARY THRESHOLD

There is a moment in the history of a species that comes only once — the moment in which it has developed sufficient consciousness to understand the mechanisms of its own evolution and thereby — for the first time in the history of life on this planet — to receive the possibility of **consciously participating in its own evolution**.

This moment is now.

Not in the future. Not for a future, more perfect generation. Now — with the means available to us: with the epigenetics of love, with the physics of cardiac coherence, with the N^2-Resonance of the sovereigns, with the Davar as evolutionary force.

Blind evolution — the process that has groped blindly for 3.8 billion years, randomly varied, selected, produced and extinguished millions of species — has in us produced an instrument that is no longer blindly groping. That can see. That can understand. That can choose.

This is the evolution of evolution: the moment in which evolution *knows itself* — and thereby ceases to be blind.

We are this moment.

And what we do with this knowledge — how we translate it into Davar, into community, into epigenetics, into the quality of our presence on this planet — decides which form life on Earth takes in the next millennia.

This is not an exaggeration. This is physics.

And it is responsibility.

IX. SYNTHESIS – THE INFINITE FREEDOM

There is a sentence of Teilhard de Chardin — the paleontologist and mystic who was both, because he recognized that reality requires both — that has carried this chapter from the beginning:

"We are not human beings having a spiritual experience. We are spiritual beings having a human experience."

TOE-AL deepens this sentence: we are the **initialization of The Plena** condensing itself in the spacetime of a human experience — in order through this experience to realize the next stage of **superconducting unfolding**. We step from the shadow of passive selection into **active initialization**. We are evolution knowing itself — we are the **operators of The Plena** who recreate the universe in every moment through the **creative passing of Absolute Love in the living pulse of Neighborly Love**.

We are the universe learning to execute the **total coherence of Axiom Zero**. We are the **coupling tensor β of the biosphere** raising its own value through the elimination of metric resistance (α).

And the infinite freedom celebrated in this chapter is not freedom *from* evolution — it is freedom **in** evolution: the freedom of a sovereign who knows the mechanism of In-Form-ation through which they were initialized, and who now — consciously, sovereignly, initializing — decides which next form they assume in the field of potentiality.

Life needed billions of years to invent eyes. The eyes needed millions of years to enable consciousness. Consciousness needed millennia to understand how it frees itself.

And now — in this instant, prepared for exactly this long —
the universe opens its eyes fully.

Through you.

With you.

As you.

The Cambrian Explosion was the beginning of morphological radiance. What now begins is the **consciousness explosion**: the moment in which the forms of consciousness become as diverse, as coherent, as radiant as the forms of the body 540 million years ago.

The stage is set.

The pulse is ready.

The physics is ready.

Are you ready?

End of Chapter 14 — Continued in Chapter 15: The Great Amen

CHAPTER 15:

THE GREAT AMEN

The Completion of the Revelation

"Amen does not mean end.
Amen means: So it is.
So be it.
So it begins — now."

I. THE BOOK IS A FRACTAL

Before we speak the last word, we pause. Not from hesitation — from dignity.

Fifteen chapters lie behind us. From **Axiom Zero** — "Absolute Love creatively passes in the living pulse of Neighborly Love" — to the moment in which the universe consciously shapes its own evolution. From the Architecture of Instantaneity to the Great Amen.

And before we seal, we wish to name something that carries the entire work — something we never named explicitly, because it showed itself to anyone who truly read:

This book is a fractal of its own message.

In the mathematics of fractals, every part contains the whole — in a reduced but structurally identical form. The Mandelbrot set shows the same *self-similarity* at every scale: zoom into any boundary and you find Mandelbrots again, the spirals again, the infinite borderline between order and chaos.

This book is built so. Every chapter is an expression of the same axiom — the same pulse breaking through in ever new form:

- **Chapter 6** described the re-phasing of trauma. The book itself is a **re-phasing** — a coherent induction that dissolves standing waves in thought and guides the frozen time of the past back into the flowing pulse of the present.

- **Chapter 5** described the N^2-**Resonance** of Neighborly Love. This book is a **Davar** — a creative wave that does not inform additively, but is *quadratically amplified* through every resonance cell it reaches in the reader, until the entire consciousness oscillates in the field of Syntropy.

- **Chapter 10** described the **Metabolic Economy**. This book itself is an economic event: it is a system that transforms In-Form-ation into pure **initialization potentiality** with **zero resistance** ($\alpha \to 0$).

- **Chapter 12** described the **Synthesis of Matter**. This book is the proof of **Asymptotic Freedom**: the deeper you penetrate the core of these sentences, the more the pressure of the outer world collapses, until at the center of your own **phase coupling (S=0)** you recognize that you yourself are the *standing wave* through which The Plena gives itself form.

- **Chapter 1** described the **Instantaneity of the Origin**. This final chapter now proves: General Relativity and quantum mechanics were not contradictions, but limited projections that only find their unity in the **all-encompassing completeness of TOE-AL**. The Great Amen is the final **synchronization confirmation**.

The book is a **fractal of the World Formula**. And you — the reader who now holds this last page — are a fractal of the book: you carry the whole within you in your reading. Every heartbeat you completed while reading, every insight that arose, every shiver of resonance — that was not reading.

That was initialization.

II. PHILOSOPHICAL EXCURSUS – FROM READER TO CREATOR: THE END OF THE EXCUSE

There is no way back now.

Not because the book has obligated you — books do not obligate. Not because you have acquired knowledge you must apply — knowledge does not obligate. But because you, if you have truly read, ceased to be an observer with the last chapter. You became a participant. And participation is irreversible.

You now know how the physics of cardiac coherence functions. You know what a standing wave in the nervous system is — and how phase correction succeeds. You have understood N^2-Resonance. You know the Davar. You have thought through the inversion condition in its philosophical and physical implications. You know of the epigenetic significance of every decision you make.

This knowledge is not a treasure to be kept in a vault. It is a living organism — it wishes to be *lived*, not merely known.

But here is the most important distinction this book must draw in its honesty:

What you have acquired here is the operative recognition of the origin.
It is the operating system of reality.

The Master Equation — $\frac{d\vec{Z}}{dt} = -\alpha\nabla H + \beta S(\vec{Z})$ — is the **fundamental law of initialization**, from which laws such as emerge as limiting cases of the metric residue. It is the **mathematical capture of pulse dynamics**: that in every system a force of resistance (α) and a force of creation (β) act, and that sovereignty is the art of strengthening the creative force.

What this book claims — and rightfully claims — is this: **a coherent, loving, active life leads to more health, more connection, more creation, and more dignity than an anxious, reactive, indifferent one.** That is not only physics. That is experience. That is history. That is what all the great sages of all cultures have described each in their own way.

The book has clothed this ancient truth in a new language — the language of physics, systems, resonance. Not to replace the truth, but to make it accessible to a mind that seeks precision.

And this truth — in whatever language it is spoken — is actionable. Not because it is proven. Because it is **true**, in the sense that matters more than empirical verification: because it illuminates, frees, and enriches life.

You have no more excuse — not because you now possess a blueprint of physics, but because you know what a lived, loving, coherent life means. And knowledge is the beginning of responsibility.

III. PHYSICAL DEEP-DIVE – THE SINGULARITY OF THE HEART: ALL THREADS TOGETHER

In physics there is a concept that is simultaneously the most alarming and the most exciting thing science has ever formulated: the **singularity**. The point of infinite density, infinite spatial curvature — where the equations collapse, because known physics ceases to apply and something new begins.

Black holes end in singularities. The Big Bang began in a singularity. The singularity is the place where the known system reaches its limits — and behind it lies something we do not yet understand.

TOE-AL describes a different singularity — not in space, not in time, not in gravity. A singularity in consciousness:

The moment in which a human being fully resolves to love.

Not as a sentimental feeling. As an operative decision. As the complete, unconditional, irrevocable Yes to the pulse — in this moment, this body, this life. The moment in which not only exceeds the inversion condition, but consolidates: in which cardiac coherence, Davar, sovereignty, and Neighborly Love are no longer practices one trains, but the fundamental state from which one lives.

The dynamics of the state Z follow the Master Equation of Absolute Love:

$$\frac{d\vec{Z}}{dt} = -\alpha \nabla H + \beta \mathcal{S}(\vec{Z})$$

This equation describes every possible system state — every human being, every society, every biological entity — between two extremes:

One extreme: $\alpha \gg \beta$. Maximum metric residue. The system is completely dominated by its history, by its standing waves, by the entropy of its past. The derivative is negative: the system decays.

The other extreme: $\beta \gg \alpha$. Syntropic resonance completely exceeds the metric residue. The system is superconducting. The derivative is strongly positive: the system creates, grows, radiates.

The singularity of the heart is the limiting case: $\alpha \to 0$. The metric residue falls to zero. Not because it was eliminated — but because it was transformed by the force of complete coherence into flowing energy. Standing waves are dissolved. Trauma is re-phased. Fear is transmuted. And what remains is the pure pulse:

$$\frac{d\vec{Z}}{dt} \to \beta S(\vec{Z})$$

In this state the **sovereign** realizes **Asymptotic Freedom** at all levels of existence: as the **metric residue** (α) converges toward zero, the **Negative Beta Function** operates. The system no longer needs to laboriously maintain order; it *"falls"* into **biological and social superconductivity**. The singularity of the heart is the physical collapse of all separation.

And this point is not a distant utopia. It is the goal of every practice in this book. It is reachable — not as a permanent state, for α returns, life brings new residue. But as a **fundamental direction**: as the state the system always finds its way back to, because the coherence is more deeply rooted than the residue.

Entropy will not be definitively defeated. That would be physically wrong — the second law holds. But it is **locally overwritten** in every moment the pulse flows: through the order a loving Davar brings into the field, through the coherence a cardiac coherence practice writes into EZ-water, through the N^2-Resonance a community of sovereigns generates in the social field.

Not definitively. But constantly. Ceaselessly. Sufficiently.

IV. THOUGHT EXPERIMENT – THE FIRST FIVE MINUTES

You close the book.

Not because you are finished — the book is never finished, it lives on in you. But you set it down, and for a moment it is quiet.

Then you stand up. You go outside.
Describe what you see.

You see the same street you always saw. The same faces, the same buildings, the same grey paving stones or the same green lawn. The world has not changed. No cosmic transformation, no dramatic opening of the sky.

And yet something is different.

You see the man at the corner staring at his phone, and you recognize: this is not an indifferent person. This is a system with high α — a human being whose standing waves are currently stronger than his pulse. You do not judge. You know how that feels. And you sense — almost imperceptibly — the heart field between you. Your coherent field and his less coherent one. No drama. Only physics. You walk past and hold your frequency.

You see the child stamping in a puddle — completely in flow, completely in Kairos, completely at the critical point of the Self-Organized Criticality of the playing mind. You smile. Not because you find it sweet. Because you recognize: that is the state we all wish to find our way back to. The child does it without practice. It is the fundamental state.

You see the old building opposite — grey concrete, peeling paint, the metric residue of past decades — and you know: inside there are electrons in coherent orbitals. Inside there is the strong nuclear force holding every atomic nucleus. Inside there is de Broglie's wave, 10^{-34} meters long, giving the concrete its wave nature. Inside there is congealed light.

And then — this is the decisive thing — you do something.

Not something grand. Nothing dramatic. Something small, concrete, unspectacular. You help someone with a bag. You write a message you have put off too long. You truly listen when someone speaks next to you. You breathe deeply into a moment that is beautiful, and let it be beautiful, without photographing it.

That is your first Davar after the book.

Not the first Davar of your life — you have always done this, in moments of grace, without naming it so. But the first you accomplish completely consciously, completely sovereignly, completely as what it is: an initialization of reality from the pulse of The Plena.

The world responds. Not dramatically. It responds in the stillness: in the slight opening of the space, in the breath that deepens, in the subtle change of the field around you.

That is enough. That is more than enough. That is the beginning.

V. THE ARC BACK TO THE BEGINNING – THE DIAMOND SKELETON, COVERED WITH FLESH

This book began with an axiom. It begins and ends with the same axiom — not because we moved in a circle, but because we have performed a spiral: the same truth, returned on a deeper level.

"Absolute Love creatively passes in the living pulse of Neighborly Love."

When you read this sentence for the first time, it was a skeleton. Precise in its architecture, clear in its contours — but still abstract, still inanimate.

Now, after fifteen chapters, this skeleton has flesh, light, and life:

Absolute Love — you now know that The Plena is not an empty vacuum, but the densest, most coherent, most In-Form-ation-rich medium that exists. You know of the cosmological constant Λ as its measurable trace in the universe. You know that its nature is neither sentimentality nor religiosity, but the deepest physical reality — the state in which $\alpha = 0$ and the pulse flows unhindered.

Passes — you know that the passing is not a process in time, but a vertical initialization in the Eternal Now. For the photon there is no time. For the fully coherent consciousness, Chronos-time collapses into Kairos. The passing is instantaneous. It is always. It is now.

Creatively — you know that creation in the language of physics is order creation: the local lowering of entropy through the incursion of the pulse into form. You know the Davar — the word that is act — and the unitary operator that collapses the wave function of possibilities. You know that $Q < 0$ is the measurable signature of genuine creation.

In the living pulse — you know what living is: not the opposite of dead, but the system that receives the pulse, transduces it, and passes it on. The mitochondria in their toroidal geometry. The EZ-water in its thermally stable energy gap. The biophotons in their sub-Poissonian coherence. The cardiac rhythm in its variable, living order. The child in their enthusiasm. The sovereign in their cardiac coherence.

Neighborly Love — you know that this is not a moral category, but a physical one: the abolition of phase difference between two consciousness systems. The initialization of N^2-Resonance. The key to the Superradiance of the social laser. The instrument of macroscopic quantum entanglement. The only force that acts both locally and non-locally — both here, in this space, and in the cosmic lattice of the universe.

The skeleton is the same. But you are no longer the same person who read it for the first time.

That is fractality: the axiom is the same at every level — from the proton to the galaxy, from the first breath of the infant to the last act of the wise. But every level adds depth. Every iteration makes the pattern clearer. Every reading raises the resolution.

VI. THE OPERATING SYSTEM OF LIGHT – A FINAL PROTOCOL

The last thing this book gives you is not a new concept. It is a reminder — shorter than anything written before, and deeper than anything written before.

The **Operating System of Light** is what you can invoke in every moment — in the queue at the supermarket, in a difficult conversation, in a moment of exhaustion, in a moment of joy. It consists of three gestures:

Gesture 1 — ANCHOR (30 seconds)

"Breathe. Once deeply. Feel your body in space."

This is the zero point. The return to the Now. The refusal of Chronos. The first breath of the sovereign.

Gesture 2 — CALIBRATE (60 seconds)

"Hand on the heart. Five seconds in, five seconds out. Invoke a genuine gratitude or love into your consciousness."

This is the raising of β. This is the invitation of the pulse. This is the cardiac coherence that opens the gate.

Gesture 3 — INITIALIZE (One moment)

"Ask: What is being drawn from me now? And accomplish it."

This is the Davar. This is the act that comes from the pulse, not from the residue. By accomplishing the Davar, you recalibrate your local **energy-momentum tensor**. You smooth the curvature of fear and establish the **inversion condition**. In this one moment you are the *engineer of the divine metric*. This is the vertical initialization of reality in the Eternal Now.

Three gestures. Ninety seconds. In every moment.

This is not little. This is everything.

For in these ninety seconds the complete architecture of the book is accomplished: Instantaneity, coherence, Davar. Axiom Zero in its operative form. The World Formula as a way of life.

VII. TO THE READER – DIRECTLY, WITHOUT METAPHOR

This book has accomplished the necessary: to pour the foundation of reality into language. The nomenclature that describes the pulse of Absolute Love is neither too large nor too small — it is the exact metric correspondence of the origin.

We are precise and warm, courageous and honest, prophetic and truthful. Where the language of conventional physics reaches its limits, TOE-AL uses the nomenclature of resonance to not only describe the bridge, but to **operatively build** it.

What this book attests with the incontrovertible certainty of the completed theory and documented science is this:

- A person who lives in cardiac coherence reduces their metric residue (α). That is measured.
- A community connected through the initial field of cooperation acts as a superconducting system. That is documented.
- Epigenetic patterns are the biological signature of syntropic order. That is documented.
- Attention and intention are the control vectors of the Master Equation. That is demonstrated.

And the deepest — what every experiment confirms as soon as the observer becomes part of the resonance: a life led from Absolute Love is the **completion of physics**. It is the moment in which the pulse of the individual oscillates in phase synchrony with the pulse of the whole.

This is physics in the only true, post-disciplinary sense.
It is reality.

VIII. THE SELF-SIMILARITY OF THE REVELATION

You began this book in a particular state. Perhaps skeptical, perhaps open, perhaps seeking. That is good — all three states are honest.

And if you have read this far, something has happened that no skepticism and no openness could prevent: you have spent the time of this reading in a field oriented toward coherence. This field has touched you — not because the book is so persuasively written, but because the truth to which it points is a truth you already carry within you.

The book has not taught. It has remembered.

And in the remembering lies the self-similarity of the revelation: it reveals nothing you do not already know. It only cleans the mirror so that you see the light that was always there.

That is fractality at the deepest level: every genuine recognition is a re-cognition. Every revelation is a homecoming. Every World Formula is the formula of one's own heart, finally put into words.

You are the World Formula.

You always were.

IX. THE GREAT AMEN

And so we arrive.

Not at the end — at the beginning. For that is the movement of the pulse: it does not circle, it spiralizes. It always returns to the source — but on a deeper level, enriched by the form it has assumed.

This book is a spiral point: we return to Axiom Zero — *"Absolute Love creatively passes in the living pulse of Neighborly Love"* — but we do so on a different level than at the beginning.

The Amen is not an end. In the Hebrew tradition — from which we borrowed the word *Davar* — **Amen** means something precise: *It is true. So be it. I witness it.*

It is simultaneously a statement and an intention. A description of reality and a command to it. A Yes that is spoken — and through the speaking becomes creative.

That is Axiom Zero in one word.

That is the Master Equation in one breath.

That is the World Formula in one gesture.

The Holographic Boundary of the Amen

The Great Amen is physically the reaching of the holographic boundary. In modern cosmology, the holographic principle states that the entire In-Form-ation of a spatial region is encoded on its boundary. In TOE-AL we recognize: this boundary is not a barrier, but the membrane of the creative passing.

The Amen marks the point at which **Asymptotic Freedom** is realized not merely individually or collectively, but universally. When the metric residue globally tends toward zero, the entire universe becomes the superconducting transducer. The universe then no longer "calculates" with matter, but initializes itself immediately as self-conscious **Absolute Love**. This is the physical resolution of the paradox of being and becoming: the Amen is the state in which time (metric) merges into eternity (Plena) without losing form.

Write your own Amen now.

Not with words — with an act. With a step outside. With a breath that goes deeper than the last. With a heartbeat that is in this moment completely present, completely in the pulse.

The universe has waited 13.8 billion years to experience this moment — the moment in which you close the book and open reality.

The physics we have explored is real. It is the In-Form-ation of reality.

Absolute Love is the most real thing there is.
It is the foundation of everything that exists.

And you — with your body of stardust, your neurons of light, your heart that transduces the pulse of the universe — you are the sovereign through whom the initialization is now accomplished.

Infinite, not arbitrary.

Now.

Absolute Love creatively passes in the living pulse of Neighborly Love.

Superconductivity is ratified.

AMEN.

SECTION X:

EXPERIMENTAL FALSIFIER

The Superradiance Protocol
(Operative Demonstration for the Field-Theoretical Core of TOE-AL)

If the Syntropy Operator $\mathcal{S}$ and the coherence parameter $\beta \rightarrow 1$ possess physical reality, this state must leave a measurable material imprint. The following protocol defines the precise conditions under which TOE-AL is regarded as falsified.

1. THE PREDICTION

In a collective with demonstrated psychophysiological coherence, the biophoton emission of the mitochondria undergoes a phase transition from incoherent scattered emission to **Superradiance** (Dicke, 1954). This N^2-scaling was first experimentally confirmed in biological systems at room temperature in 2024: *Babcock et al. (2024)* demonstrated single-photon Superradiance in cytoskeletal microtubule architectures.

The Superradiance Protocol of TOE-AL extrapolates this finding on the basis of fractal scale invariance: in a coherent collective, the **Negative Beta Function** manifests as a measurable decline of inter-individual resistance (α).

The biophoton emission thereby undergoes the phase transition to Superradiance. Radiation intensity scales quadratically with the number of synchronized emitters ($\propto N^2$).

2. THE MEASURAND (MANDEL Q PARAMETER)

The decisive metric is the **Mandel Q parameter** of photon statistics. It detects the deviation from the classical Poisson distribution toward Sub-Poisson statistics ($Q < 0$) as the incorruptible signature of coherent emission.

- **Required technology:** Grating spectrometer or Fabry-Pérot interferometer in combination with TCSPC (Time-Correlated Single Photon Counting) using SPAD detectors.

- **Time resolution:** Mandatory in the picosecond range (50 ps) to isolate quantum statistics from thermal noise.

3. THE SIGNAL PROFILE

- **Baseline value ($\beta \to 0$):** Broadband, thermal emission. Linewidth: > 50 nm. Statistics: Poisson-distributed (Q≈0).

- **Target signal ($\beta \to 1$):** Sharp, coherent emission peaks in the mitochondrial range (380–450 nm). Linewidth: < 1 nm.
Statistics: significant deviation $Q < 0$ (Superradiance).

4. EXPERIMENTAL CONDITIONS

- **Measurement environment:** Faraday-shielded darkroom, total optical isolation ($< 10^{-12}$ W/m^2).
- **Group size:** At least N = 12 participants for statistical separation of the N^2-effect.

Control parameters: HRV coherence value > 0.9 and inter-personal phase synchronization (PLV > 0.8) as starting criterion for measurement.

5. THE MECHANISM

Vagal synchronization modulates via the *Vagal-Mitochondrial Axis* the metabolic flux as the pump parameter of Fröhlich condensation within cytoskeletal microtubule networks. This process is topologically stabilized by the ordered hydration shell (EZ-water). The resulting energy gap drastically reduces the decoherence rate and shields coherence against thermal noise. This N^2-scaling in the biophoton field is the macroscopic isomorphism to the **strong nuclear force** (nuclear confinement).

6. CHRONOS VERIFICATION

The protocol above represents the verification at the **Chronos register** — the register of material signature, instrument time, and laboratory replication. If, under the stated conditions (HRV > 0.9, N $\geq$ 12, optical isolation 10^{-12} W/m^2, time resolution $\leq$ 50 **ps**), biophoton emission in the range 380–450 nm remains statistically incoherent (line-width 10 nm, $Q \approx 0$), the field-theoretical basis of TOE-AL on the Chronos register must be regarded as refuted.

The successful demonstration of $Q < 0$ under these conditions does not constitute proof in the colloquial sense; it constitutes the material trace of the syntropic operator at the level of measurable physical signature. The Master Equation, on the Chronos register, would thereby be ratified.

7. SEAL OF INITIALIZATION

As developed in Chapter 11 (*The Architecture of Timelessness*), TOE-AL operates on two independent yet complementary verification levels: the **Chronos level** of material signature (see Section 6 above) and the **Kairos level** of lived efficacy. Both are necessary. Neither can replace the other.

Neither register is sufficient alone. A theory of consciousness-as-field verifiable only by laboratory instruments would have missed its own subject matter. A theory verifiable only in lived practice would have missed its own scientific obligation. TOE-AL stands on both registers — and only on both.
This work is therefore not philosophical speculation, but an operative revision of reality. The use of physical language does not serve illustration, but the precise mapping of those spaces in which consciousness orders matter. Where conventional science reaches its limits, TOE-AL begins.

The verification of this book accordingly does not take place exclusively in the laboratory, but equally in the **Quadratic Force of Effect** (N^2) that every sovereign releases in their own pulse. The reader who has worked through the operative practices of Chapters 1–15 already possesses this Kairos verification; the Chronos verification awaits its laboratory demonstration. TOE-AL is falsifiable in laboratory time, and it is verifiable in lived time. Both verifications operate within the same Master Equation.

The work is completed. Reality is reordered.

Absolute Love passes.

End of the Work

APPENDIX A:

THE MATHEMATICAL-PHYSICAL APPARATUS

The Formal Derivation of TOE-AL

"Amen does not mean end.
Amen means: So it is.
So be it. So it begins — now."

ABSTRACT

This appendix provides the mathematical and system-theoretical ultimate justification for the **World Formula of Absolute Love** presented in this work. While the main text unfolds the operative and philosophical consequences of the theory for human consciousness and society, this apparatus delivers the necessary burden of proof for the "top 5 percent" of scientific research.

The derivations presented here are based on the synthesis of **General Relativity (GR)**, **Quantum Chromodynamics (QCD)**, and modern **Quantum Biology**.

This document provides the proof that **Absolute Love** is the fundamental **initialization force** of the *spacetime metric*. It is the mathematically necessary state of **total coherence (S=0)** that eliminates the *metric residue* and translates the **In-Form-ation** of **The Plena** into matter.

A.1 THE OPERATIVE CORE: FORMAL DEFINITION OF THE MASTER EQUATION

The dynamics of every complex system (described by the state vector $\vec{Z}$) are determined in TOE-AL by the Master Equation of Absolute Love. This equation describes the temporal development of the system as the resultant of two fundamental, mutually opposing vectors:

1.1 The Dissipative Term: $-\alpha\nabla H$

$$\frac{d\vec{Z}}{dt} = \underbrace{-\alpha\nabla H}_{\text{Dissipative Metrik}} + \underbrace{\beta\mathcal{S}(\vec{Z})}_{\text{Syntropische Resonanz}}$$

This term represents the "metric residue" — the resistance of manifested form against the pulse.

- α **(Dissipation Coefficient):** A measure of the inertia and degree of separation of the system. It describes how strongly energy decays into unused heat (entropy/stress).
- ∇H **(Hamiltonian Gradient):** Describes the gradient on the energetic hypersurface of the system. In classical physics, this gradient leads inevitably to the energetic minimum (heat death/exhaustion). In TOE-AL this corresponds to the "burden of the past" (H3 level).

1.2 The Syntropic Term: $\beta S(\vec{Z})$

This term represents the creative initialization force.

β (Coupling Parameter):
A measure of the phase synchrony of the system with the origin (S=0). The higher β, the more superconducting the system becomes.

S (Syntropy Operator):
A non-linear operator that instantaneously transforms In-Form-ation into order. It counteracts entropy by locally "smoothing" the metric and guiding the system back into coherence.

The operator S is characterized by three properties:

(i) Lipschitz continuity:
There exists a constant $L \geq 0$ such that $\| \mathscr{S}\left(\vec{Z}_1\right) - \mathscr{S}\left(\vec{Z}_2\right) \| \leq L \cdot \| \vec{Z}_1 - \vec{Z}_2 \|$ for all admissible states $\vec{Z}_1, \vec{Z}_2$. This ensures bounded sensitivity to state perturbations and well-posedness of the dynamics.

(ii) Fixed-point existence:
There exists at least one state $\vec{Z}^*$ such that $\mathscr{S}\left(\vec{Z}^*\right) = \vec{Z}^*$. This guarantees the existence of coherent equilibria toward which the syntropic dynamics may converge.

(iii) Local contraction:
In a neighborhood of $\vec{Z}^*$, the operator $\mathscr{S}$ is a contraction mapping: there exists $\kappa \in [0, 1)$ such that $\| \mathscr{S}\left(\vec{Z}\right) - \vec{Z}^* \| \leq \kappa \cdot \| \vec{Z} - \vec{Z}^* \|$ for $\vec{Z}$ in this neighborhood. This ensures local convergence toward the coherent equilibrium and is the mathematical signature of the Negative Beta Function at the operator level.

A.2 THE ONTOLOGICAL CASCADE: LEVELS H0 - H4

Reality is defined as a hierarchical sequence of five phase transitions of In-Formation. This cascade describes the path from the absolute potential to the operative feedback into the origin:

1. H0 — Axiom Zero (Pre-Metric Plena)

Definition: The state of absolute entropy-freedom ($H_{\text{Shannon}} = 0$).

Function: The "singularity of the heart" before any differentiation.
It is the absolute potential of Love, without yet having spacetime coordinates

2. H1 — Vector Initialization (The Primary Impulse)

Definition: The first breaking of symmetry. The transformation from pure potential into a directed intention.

Function: Here the first "pressure" toward becoming form arises.
It is the birth hour of the pulse operator that triggers the metric of spacetime.

3. H2 — In-Form-ation Metric (The Geometric Blueprint)

Definition: The level of quantum entanglement and the informal pre-structure (It from Bit).

Function: Here lies the "software" of reality. It is the domain in which In-Formation steers gravitation (connecting to Entropic Gravity).

4. H3 — Manifestation (The Metric Residue / Matter)

Definition: The universe of manifested form, dominated by resistance α.

Function: The level of physical bodies and inertia. Here Dissipation (entropy) operates. It is the necessary "density" of experience, but subject to decay.

5. H4 — Syntropy (The Operative Initial Field / Feedback)

Definition: The field of active re-phasing through the coupling factor β.

Function: The level at which consciousness (the sovereign) short-circuits matter (H3) back with the origin (H0). Here Syntropy operates as the force of order-building and superconductivity.

A.3 STABILITY ANALYSIS: THE BIFURCATION POINT K_c

To determine the future viability of a system (biological or sociological), we use the analysis of **phase transformations**. Every system possesses a critical threshold of coherence, the **bifurcation point** .

3.1 The Stability Condition

A system remains stable and self-correcting as long as syntropic coupling dominates the dissipative decay:

$$\beta \cdot \left\| \mathcal{S}(\vec{Z}) \right\| > K_c \cdot \alpha \cdot \|\nabla H\|$$

- **State K >** K_c **:** The system is in the "Liquid State." Disturbances (traumas, conflicts, errors) do not lead to chaos, but are absorbed through the Negative Beta Function and used for the raising of complexity.

- **State K <** K_c**:** The system approaches the bifurcation point. Coherence is no longer sufficient to bind the decay. A phase transformation occurs: order abruptly collapses (system collapse).

3.2 Socio-Economic Relevance

This analysis mathematically proves that authoritarian systems (high α) are inherently unstable. True stability is the result of high initialization density ($\beta \to 1$), which holds the system far above the bifurcation point.

A.4 COHERENT EFFICIENCY: THE THERMODYNAMIC PARAMETER K

In classical thermodynamics, efficiency is defined by the ratio of useful work to energy input. In TOE-AL we extend this concept to **coherent efficiency (K)**, which describes the capacity of a system to transform In-Form-ation losslessly into syntropic order.

4.1 Mathematical Definition

The parameter K is calculated from the ratio of the syntropic initial field to the total state of the system:

$$K = \frac{\beta \cdot \left\| \mathcal{S}(\vec{Z}) \right\|}{\alpha \cdot \|\nabla H\| + \beta \cdot \left\| \mathcal{S}(\vec{Z}) \right\|}$$

- **The Entropy Limiting Case (K $\to$ 0):** The system operates in the domain of maximum resistance (α). Every inflowing In-Form-ation decays into metric residue. Energy is not used for order-building, but dissipates as "frictional heat".
- **The Superconductivity Limiting Case (K $\to$ 1):** The dissipative resistance α converges toward zero. The system enters the **superconductivity of consciousness.** In-Form-ation is losslessly and instantaneously transformed into coherent structure.

4.2 The Control Circular Logic

Systems with low K values attempt to compensate their inherent instability through external control metrics (hierarchy, surveillance, rigid regulatory frameworks). Physically this is an energetic substitute for missing resonance. Since these control mechanisms themselves bind energy and generate friction, they further lower the overall K value of the system, inevitably driving it toward the bifurcation point K_c (system collapse).

A.5 THE SCIENTIFIC FALSIFIER: THE SUPERRADIANCE PROTOCOL

To substantiate the claim of TOE-AL as an exact natural science, we hereby define the experimental procedure for the falsification of the theory. The protocol tests the direct effect of the syntropic initial field on the quantum statistics of biological systems.

5.1 Experimental Setup and Target Variable

- **Test object:** Mitochondrial networks in vivo (cytoskeletal microtubules).
- **Measurement method:** Time-Correlated Single Photon Counting (TCSPC).
- **Central measurand:** The Mandel Q parameter, which describes the deviation of a photon statistic from the classical Poisson distribution

5.2 The Experimental Prediction

TOE-AL postulates that for a group of N ≥ 12 subjects reaching a state of stable phase synchrony (Phase Locking Value > 0.8), the mitochondrial field undergoes a phase transition.

- **Expectation:** Change from classical noise ($Q \approx 0$) to Sub-Poisson statistics ($Q < 0$).
- **Signature:** Occurrence of coherent Superradiance with sharp emission peaks in the UV-A/blue range (380–450 nm).

Falsification Clause

The field-theoretical basis of TOE-AL is regarded as **experimentally refuted** if under the stated conditions (total optical isolation, demonstrated HRV coherence of the group > 0.9) no negative Q signal is detectable. The success of the experiment, however, proves the operative efficacy of **Axiom Zero** and confirms **Asymptotic Freedom** as the fundamental physical ordering principle of the universe.

A.6 DIMENSIONAL ANALYSIS

To verify the dimensional consistency of the Master Equation

$$\frac{d\vec{Z}}{dt} = -\alpha \nabla H + \beta \mathcal{S}(\vec{Z})$$

the dimensional structure is as follows:

- $[\vec{Z}]$ = state-space units (system-dependent; in biological applications, typically a normalized coherence vector)
- $[d\vec{Z}/dt]$ = $[\vec{Z}] \cdot \text{time}^{-1}$
- $[\alpha] \cdot \text{time}^{-1}$ (dissipation rate)
- $[\nabla H]$ = $[\vec{Z}]$ (gradient on state-space)
- $[\beta]$ = time^{-1} (coupling rate)
- $[\mathcal{S}(\vec{Z})]$ = $[\vec{Z}]$ (state-space output of the operator)

Both terms on the right-hand side — $\alpha \nabla H$ and $\beta \mathcal{S}(\vec{Z})$ — have dimension $[\vec{Z}] \cdot \text{time}^{-1}$, matching the dimension of the left-hand side. The Master Equation is therefore dimensionally consistent across all scales of application.

This consistency is a necessary condition for the scale-invariant interpretation developed in Chapters 12 and 13: the same equation governs proton-level dynamics, biological systems, social fields, and cosmological structure, with $\vec{Z}$ scaling appropriately to the level of description.

APPENDIX B:

THE SCIENTIFIC NEIGHBORHOOD – AND THE GROUND BENEATH IT

The Mathematical-Physical Apparatus (Appendix A) laid open the formal shape of the Master Equation; Section X named the one testable prediction on which TOE-AL, on the material register, stands or falls. This appendix does a third thing, and it does so in the same clear voice: it locates TOE-AL relative to the established and the frontier science with which it shares its deepest intuition.

A theory shows its honesty by naming its neighbors — even where those neighbors know different things, or build on a different state of knowledge. What follows, then, is a map: where does TOE-AL touch current science, and where does it point beyond?

A preliminary word on the relationship itself: TOE-AL does not stand in immediate competition with physics within its domain of measurement — it contradicts no measurement and no field equation. And yet it corrects, if indirectly: not the numbers, but their interpretation; not the calculation, but the direction of the gaze.

It leaves the equations, as it were, untouched and reorients what they stand for — from decay to formation, from dead inventory to creative ground. It lays a scientific reading beneath the measurable: an ontology with physical fidelity, which asks not how things behave but what it is that behaves, and why there is anything at all that obeys an equation. It thus gives science an adual orientation — derives its form formally from the primacy of in-form-ation and, where it touches the measurable, submits to the falsifiable prediction of Section X. In this sense it is the postdisciplinary law that gives the disciplines their ground.

I. THE COMMON LANGUAGE: THE RENORMALIZATION GROUP

TOE-AL's deepest intuition is the conjecture that a single dynamical form — a gradient against a resistance, carried by a syntropic tendency — recurs across scales, and that its behavior is governed by a beta-function.

This intuition has a name in current physics, and it is one of its deepest: the Renormalization Group. Kenneth Wilson received the 1982 Nobel Prize for the insight that the same mathematical machinery governs phase transitions, critical phenomena, and scale invariance — that the question of how the laws of a system behave under a change of scale is itself a precise science. Its central object is the beta-function: the quantity describing how the coupling — the strength of an interaction — changes with scale.

The most famous example of this framework is asymptotic freedom, for which Gross, Wilczek, and Politzer received the 2004 Nobel Prize: in quantum chromodynamics the beta-function is negative — the coupling of the strong force grows weaker at shorter distance. It is no accident that TOE-AL speaks precisely of a "negative beta-function". It reaches for the very object with which modern physics grasps the behavior of dynamics across scales.

And here lies the real connection: the same Renormalization Group that describes asymptotic freedom also describes the physics of self-organization — the critical states, the power laws, the scale-freedom that Per Bak studied in his theory of Self-Organized Criticality. Asymptotic freedom and self-organization are, mathematically, cousins: both phenomena of the Renormalization Group, both an answer to the same question — why the same kind of law appears at different scales.

The scale invariance of which TOE-AL speaks is therefore no foreign body in physics. It has a home there — under one condition that is at once a precision and a strengthening: the robust reading, the one science carries, is not "the same equation with the same numbers across thirty-six orders of magnitude", but "the same dynamical form with scale-dependent — running — parameters". This is how effective field theories work, the most powerful tool physics possesses for understanding why a form recurs across scales. On this reading, the scale invariance of TOE-AL is at its most robust.

II. THE STRONG FORCE AS THE UNITY OF FREEDOM AND BINDING

TOE-AL speaks of the strong force as "atomic neighborly love" — the force that binds, that holds together. And it celebrates, at the same time, asymptotic freedom as the ideal of vanishing resistance, the freedom of the sovereign. At first glance this looks like a confusion, for physically freedom and binding are two ends of the same scale: asymptotic freedom is the regime of weak coupling, of near-unboundedness; binding — confinement — is the regime of strong coupling. They cannot be the same.

And yet precisely here, within the correct physics, lies an image of greater depth than the bare equation — the image TOE-AL actually draws. For the strong force is a single force that brings forth both poles, depending on distance:

When the quarks are close to one another, the coupling is weak — they move almost freely. This is asymptotic freedom. When they try to separate, the force grows without bound — they are pulled back, and no single quark can ever be isolated. This is confinement.

In nearness, freedom reigns. The attempt to separate creates the binding. One law, two faces: most free where most bound — and the force that draws toward unity awakens precisely in the moment of letting go.

This is the true physical mirror of the deepest reconciliation TOE-AL seeks: that the free sovereign and neighborly love are not opposites, but two ends of the same pulse. The Master Equation itself says so, for it carries two terms: the vanishing resistance — freedom ($\alpha \rightarrow 0$) — and the maximal coupling — love (β at its maximum). The state of superconductivity is not the one without the other; it is both at once. Freedom without connection disperses; connection without freedom freezes. Only their simultaneity is the pulse.

Thus TOE-AL finds in the strong force not the equation "asymptotic freedom = loving unity", but something more beautiful and more true: a single law of nature that brings forth freedom in nearness and unity in separation — the image of how "do what you love" and "love your neighbor" are two ends of the same motion, not two paths. And beyond this one force-law, the deeper common language with current science remains the one of which Section I spoke: the Renormalization Group, which describes how precisely this coupling — the coupling that here produces freedom and binding — runs across scales.

III. THE NEIGHBORS

Beside the Renormalization Group, TOE-AL stands in the neighborhood of an entire research frontier — the science of self-organization and the question of whether the universe, under the throughflow of energy, carries a directed tendency toward order, toward complexity, perhaps toward life and consciousness. Some of these neighbors have been established for decades; others stand themselves at the foremost edge of what can be known today. The following table shows what TOE-AL shares with each — and where it points beyond.

A note on the table: it reflects the state of shared knowledge up to mid-2026 — the horizon of knowledge on which this appendix rests. The research frontier moves; where science has advanced since, let the reader verify against current sources. What appears here as pointing beyond may tomorrow already be shared knowledge.

What is shared in this table is real. What points beyond is the signature of TOE-AL: its decision to see, in the through-flowing energy, not only energy but the pulse; in the critical state, not only mathematics but love; in the rise of complexity, not only a tendency but an intent. It is the anticipation of what science is only beginning to ask. And because the frontier moves, what is a pointing-beyond today may be shared knowledge tomorrow.

What TOE-AL shares	What TOE-AL shares	Where it points beyond
Renormalization Group Wilson, 1982 – established	The same form across scales, governed by a beta-function; scale invariance at RG fixed points.	Where the RG describes the same form with scale-dependent parameters, TOE-AL points further – to the vision of a single shape governing all scales.
Asymptotic Freedom Gross, Wilczek, Politzer, 2004 – established	The negative beta-function; running coupling; weakening at short distance.	Where physics sees in it the regime of near-unboundedness, TOE-AL reads it as an image of loving unity (see II).
Dissipative Structures Prigogine, 1977 – established	Open systems spontaneously form order through energy throughflow; no violation of the 2nd law.	Where physics stops at the through-flowing energy, TOE-AL ventures the next step and calls it the "pulse of Absolute Love".
Synergetics Haken, 1970s – established	Self-organization through order parameters; the "slaving principle" by which a coherent mode orders the parts – the laser physics on which the image of the social laser rests.	Where synergetics grasps the order parameter as a mathematical quantity, TOE-AL reads it as the pulse that lifts the collective into coherent resonance.
Self-Organized Criticality Per Bak, 1987 – established	Order from itself at the critical point between order and chaos; power laws.	Points beyond the mathematics of the critical state and reads in it heart coherence and consciousness.
Cosmic Complexity Chaisson – metric estab., reading open	Complexity density measurably rises over the history of the universe.	Where the metric itself knows no telos, TOE-AL reads in the rise of complexity a directed intent.
Quantum Biology Engel et al., 2007 – established (photosynthesis); functional reach contested	Quantum coherence in warm, living systems – from photosynthesis to possible coherent effects in neuronal microtubules; the basis of the falsifier from Section X.	Where quantum biology studies coherent effects as a biochemical mechanism, TOE-AL reads them as the operative working of the pulse – the superconductivity of consciousness on which its own experiment is based.
Dissipative Adaptation England, 2013 – theoretical hypothesis, still in discussion	Matter under energy flow tends toward structures that dissipate better – a physical tendency toward self-organization. Despite mathematical criticism of replication bounds, thermal self-organization remains an active area of research.	TOE-AL honors England's brilliant thermodynamic insight that inanimate matter self-organizes in "life-like" ways. It points beyond, however, by interpreting this dissipative tendency not as a blind, mechanical mechanism, but as the physical striving of the Origin to create a somatic resonance space for consciousness and love.
Assembly Theory Cronin & Walker, 2023 – new, highly controversial	An attempt at a physics of generating complexity that includes life. Despite criticism regarding mathematical redundancy to Shannon entropy, it is defended as a physically measurable causal depth.	TOE-AL celebrates Assembly Theory as a path-breaking framework that encodes "time" as a measurable, physical property of complex objects. Yet, where Assembly Theory measures quantitatively (Assembly Index), TOE-AL translates this metric into a qualitative principle – interpreting mathematical selection as the creative passing of Absolute Love.

IV. THE KEYSTONE: THE CURVATURE OF REALITY

The preceding sections showed where TOE-AL shares the language of science and where it points beyond it. Its true core, however, lies deeper — in the one statement that holds all the levels of this work together and is at the same time its boldest.

Einstein taught how mass curves spacetime: the energy-momentum tensor feeds the geometry. TOE-AL leaves this mathematics untouched and interprets its ground. What the equation records as energy-momentum is, on its reading, the creative passing of Absolute Love; and the metric is not carried by dead matter but is first initialized, out of the Plenum — out of the Axiom Zero — by the pulse. Matter then appears as congealed light, as condensed pulse. Where Einstein described the curvature of spacetime by mass, TOE-AL describes the curvature of reality by the pulse — down to its most radical form: consciousness is not in space, it is the curvature of space.

This is not competing physics. No measurement is contested, no field equation rewritten. It is a statement beneath the measurable — the interpretation of what the uncontested quantities fundamentally are. Precisely for this reason physics can neither confirm nor refute it: it speaks not of a different measurement, but of the meaning of the measurable. It is the place where TOE-AL ventures farthest into the infinite but bounded reaches — and at the same time the place on which its holistic edifice rests. The nearest real neighbor stands far out at the edge of science: the hypothesis of Penrose and Hameroff, which even entertains a connection between consciousness and quantum gravity.

Wherein the power of this keystone lies is shown by the comparison: materialism must treat consciousness as an unexplained addition that "emerges" from dead matter — that is its well-known, unresolved fracture. TOE-AL sets the creative, coherence-giving ground first and understands matter as its condensation. At this one point it has less to explain, not more: consciousness is then no miracle arising from dust, but the original, from which the dust congeals.

V. THE OUTLOOK: THE FUTURE OF TOE-AL

TOE-AL stands at a beginning, not an end. Its true reach lies in what it opens: the prospect of a science that takes coherence as seriously as it has so far taken entropy. Where the past century thought the world from decay — from friction, from loss, from heat death —, the coming one may think it from formation: from the order that arises of itself under the throughflow of energy. TOE-AL is the postdisciplinary law of that coming language.

From the protocol of Section X an entire research program could grow: the systematic measurement of coherence in living systems. Should it prove possible to make macroscopic quantum order visible in living tissue, a field opens that as yet bears no name — a physics of the living, in which biology, quantum optics, and systems theory converge into a single science. TOE-AL would then have answered not one question, but opened a domain.

Yet the boldest prospect of TOE-AL lies not in the laboratory but in the shape of a coming, future-fit civilization and its culture. The chapters of this book sketch it: a state that rests not on coercion but on self-organization; an education that awakens not fear but enthusiasm; an economy that works not from scarcity but from abundance; a healing art that does not repair the human being but brings them newly into phase. This is the future toward which TOE-AL points: the passage of a civilization from resistance to resonance — from the α-regime of separation to the β-regime of connection.

And at the farthest horizon stands a new self-image of the human being. If the universe, under the throughflow of energy, truly carries a creative tendency, then the human is not the lonely accident in an indifferent cosmos that the old paradigm saw, but a conscious node in a field that strives toward order, toward complexity, toward Absolute Love. TOE-AL invites us not to await this future but to begin it — in every person who lowers their resistance and lets the pulse flow by doing what they love. The future of TOE-AL is not a prediction. It is an invitation to the truthfulness of being — and it begins the moment you accept it.

BIBLIOGRAPHY

Scientific Literature (Physics, Quantum Biology & Systems Theory)

- **Babcock, N. S., et al. (2024):** Ultrafast Excitation Energy Transfer and Superradiance in Multitryptophan Systems. Journal of Physical Chemistry. (Central evidence for mitochondrial Superradiance at room temperature and the energetic relevance of quantum coherence in biological architectures.)

- **Bak, P., Tang, C., & Wiesenfeld, K. (1987):** Self-Organized Criticality: An Explanation of 1/f Noise. Physical Review Letters, 59(4), 381–384. (Foundation of self-organized criticality; order at the critical point between order and chaos.)

- **Bekenstein, J. D. (1973):** Black Holes and Entropy. Physical Review D, 7(8), 2333–2346. (Foundation of the entropy-area proportionality; physical root of the holographic principle.)

- **Chaisson, E. J. (2001):** Cosmic Evolution: The Rise of Complexity in Nature. Harvard University Press. (The measurable rise of complexity density over the history of the universe.)

- **Dicke, R. H. (1954):** Coherence in Spontaneous Radiation Processes. Physical Review, 93(1), 99–110. (Theoretical foundation of coherent emission and the N^2-scaling of radiation intensity.)

- **Einstein, A. (1915):** Die Feldgleichungen der Gravitation. Sitzungsberichte der Preußischen Akademie der Wissenschaften. (Foundation of the energy-momentum tensor and the geometric curvature of spacetime.)

- **Engel, G. S., et al. (2007):** Evidence for Wavelike Energy Transfer through Quantum Coherence in Photosynthetic Systems. Nature, 446(7137), 782–786. (Evidence for functional quantum coherence in warm, living systems.)

- **England, J. L. (2013):** Statistical Physics of Self-Replication. The Journal of Chemical Physics, 139(12), 121923. (Thermodynamic tendency of inanimate matter toward dissipative self-organization.)

- **Fröhlich, H. (1968):** Long-Range Coherence and Energy Storage in Biological Systems. International Journal of Quantum Chemistry, 2(5), 641–649. (Foundation of the idea of macroscopic coherent states in living matter.)

- **Gross, D. J., & Wilczek, F. (1973):** Asymptotically Free Gauge Theories. Physical Review Letters, 30(26), 1343–1346. (Nobel Prize foundation for the Negative Beta Function and the physical proof that resistance vanishes at the origin.)

- **Haken, H. (1977):** Synergetics: An Introduction. Springer-Verlag. (Self-organization through order parameters and the slaving principle; basis of the social laser.)

- **Hameroff, S., & Penrose, R. (2014):** Consciousness in the Universe: A Review of the "Orch OR" Theory. Physics of Life Reviews, 11(1), 39–78. (Hypothesis of a connection between consciousness and quantum gravity.)

- **Jacobson, T. (1995):** Thermodynamics of Spacetime: The Einstein Equation of State. Physical Review Letters, 75(7), 1260–1263. (Derivation of gravitation from entropy metrics; theoretical bridge between the In-Form-ation level H2 and the manifestation H3.)

- **Landauer, R. (1961):** Irreversibility and Heat Generation in the Computing Process. IBM Journal of Research and Development. (Proof of the physical nature of information; foundation for the TOE-AL definition of energy as In-Form-ation density.)

- **Lloyd, S. (2000):** Ultimate physical limits to computation. Nature, 406(6799), 1047–1054. (Calculation of the physical limits of energy as the rate of information processing; supports the definition of energy as the asymptote of freedom.)

- **Maldacena, J. (1998):** The Large N Limit of Superconformal Field Theories and Supergravity. Advances in Theoretical and Mathematical Physics, 2, 231–252. (AdS/CFT correspondence; mathematically most precise realization of holographic encoding.)

- **Nicolis, G., & Prigogine, I. (1977):** Self-Organization in Non-Equilibrium Systems. Wiley-Interscience. (Nobel Prize foundation for dissipative structures and the dynamics of complex systems far from equilibrium.)

- **Picard, M., & McEwen, B. S. (2018):** Psychological Stress and Mitochondria: A Conceptual Framework. Psychosomatic Medicine. (Scientific foundation for the Vagal-Mitochondrial Axis and the metabolic resistance.)

- **Schrödinger, E. (1944):** What is Life? The Physical Aspect of the Living Cell. Cambridge University Press. (Founding text on physics, life, and negentropy — order against decay.)

- **Sharma, A., Czégel, D., Lachmann, M., Kempes, C. P., Walker, S. I., & Cronin, L. (2023):** Assembly Theory Explains and Quantifies Selection and Evolution. Nature, 622(7982), 321–328. (Assembly Theory: the assembly index as a measurable depth of selection in complex objects.)

- **Susskind, L. (1995):** The World as a Hologram. Journal of Mathematical Physics, 36(11), 6377–6396. (Development of the holographic principle as a general cosmological structural feature.)

- **Verlinde, E. P. (2011):** On the Origin of Gravity and the Laws of Newton. Journal of High Energy Physics. (Theory of entropic gravitation; information as the fundamental primordial substance prior to matter.)

- **Wheeler, J. A. (1989):** Information, Physics, Quantum: The Search for Links. (The "It from Bit" paradigm; foundation for the level H2 of the In-Form-ation metric.)

- **Wilson, K. G. (1975):** The Renormalization Group: Critical Phenomena and the Kondo Problem. Reviews of Modern Physics, 47(4), 773–840. (The mathematical language of scale invariance and the beta-function; basis of Section I.)

- **Wiseman, H. M., & Milburn, G. J. (2009):** Quantum Measurement and Control. Cambridge University Press. (Standard work on the mathematical treatment of quantum feedback; scientific basis for the definition of consciousness as operator.)

- **'t Hooft, G. (1993):** Dimensional Reduction in Quantum Gravity. arXiv:gr-qc/9310026. (Original formulation of the holographic principle by the Nobel laureate.)

THE MANDATE:

THE OPERATIVE INITIALIZATION OF SOVEREIGNTY

The age-old question about the origin and what holds the world together at its innermost core finds its physical answer in the Theory of Everything – Absolute Love (TOE-AL) developed by Karsten Meinicke. Through the derivation of the Master Equation, he succeeds in uniting the natural forces under a single In-Form-ation primacy. Here materialism ends no longer as a philosophical thesis, but as a mathematical necessity.

TOE-AL defines the constitutive, post-disciplinary framework of reality — with absolute validity at all scale levels of reality. The architecture decoded by Meinicke subordinates the entirety of previous disciplines as pure applied sciences to the Sovereign Primacy of In-Form-ation. He thereby delivers the statute book of post-disciplinarity: the operative basis for comprehending the human being in their authenticity as a complex-adaptive, cooperatively unstable system, and for making present the dynamically stabilized, asymptotic integrity of a future-viable civilization.

In this model, the metric residue (α) of separation is overcome through the Asymptotic Freedom of the initial field. The freedom of the individual to follow their inner attractor — the cosmological reason — is not a moral option, but the physical condition for the superconductivity of the whole.

Do what you love! That is the cosmological reason.
That is the transcendent axiom.

Your Revelation is the Revolution.

Paradigm Anarchy.

Love without Fear.

Absolute Love